Global Productions
Labor in the Making of the
"Information Society"

The Hampton Press Communication Series
International Communication
Richard C. Vincent, supervisory editor

Good-Bye, Gweilo: Public Opinion and the 1997 Problem in Hong Kong
L. Erwin Atwood and Ann Marie Major

Democratizing Communication?: Comparative Perspectives on
Information and Power
Mashoed Bailie and Dwayne Winseck, eds.

Global Productions: Labor in the Making of the "Information Society"
Gerald Sussman and John A. Lent, eds.

Reconvergence: A Political Economy of Telecommunications in Canada
Dwayne Winseck

forthcoming

Towards Equity in Global Communication: MacBride Report Update
Richard Vincent, Kaarle Nordenstreng, and Michael Traber, eds.

Political Economy of Media and Culture in Peripheral Singapore
Kokkeong Wong

Global Productions
Labor in the Making of the
"Information Society"

edited by

Gerald Sussman
Portland State University

John A. Lent
Temple University

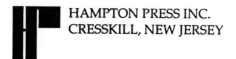

HAMPTON PRESS INC.
CRESSKILL, NEW JERSEY

Printed in the United States of America

Library of Congress Cataloging-in-Publication Data

Global productions : labor in the making of the "information society"
/ edited by Gerald Sussman, John A. Lent.
     p.   cm. -- (The Hampton Press communication series)
    Includes bibliographical references and indexes.
    ISBN 1-57273-171-0. -- ISBN 1-57273-172-9 (pbk.)
    1. High technology industries--Employees--Supply and demand--
Case studies.   I. Sussman, Gerald. II. Lent, John A. III. Series.
HD5718.H54G58 1998
331.11'92139--dc21   98-16989
                                              CIP

Cover design: Tom Phon Graphics and Gerald Sussman

Hampton Press, Inc.
23 Broadway
Cresskill, NJ 07626

For the communication and information workers of the world

Contents

Prologue
The Gods and the Mortals

David F. Noble

Global Productions is a study in contrast. It juxtaposes two starkly different domains that might be called a world of gods and a world of mortals. At the same time, the book shows how these separate worlds are linked, as actually but two sides of a single realm, and how each defines and is dependent upon the other. For now, it appears, the gods are on the top side, the mortals on the bottom, but there is not telling how long that will last.

Life on the upper side of this realm is certainly familiar, since it is depicted daily on millions of screens, newspaper and magazine pages, and billboards throughout the world. The gods, it would seem, are ubiquitous. The universality of the god-life is the single most important image of the so-called information age, not only because it purposefully portrays the fulfillment of every consumer's dream, but because it makes it seem as if there are no mortals around anymore. Ethereal in their electronic cyber-existence, armed with their computers, telecommunication networks, satellites, modems, faxes, e-mail, and the Internet, the god are everywhere, instantaneously. They know everything. Their companies are transnational but they are transcendent.

They have overcome their mortality, the constraints of space and time, the particulars of place and the moment. Thus they exult in their godliness, their virtual divinity, the culmination of a thousand years of earnest expectation.

But there are still mortals around, on the bottom side of the realm, outnumbering the gods millions-to-one. Unheard and unseen on top-side screens, these are the people who make, operate, and repair the things that make the gods gods — the electronic components, the computer assemblies, the telecommunication links, the data entry. But they cannot themselves afford the things they make, operate, and repair, so they remain mortals, their drudgery and desperation the unspoken and unsung human realities underlying topside delusions of divinity. They inhabit not merely a different place but a different dimension. They are not transcendent, but earth-bound, tied to the particulars of place and time by assembly lines and punch-clocks, surveillance and monitoring systems, managerial authority and the detail division of labor. They exist not in cyberspace but in places like Barbados, China, the Dominican Republic, Hong Kong, India, Ireland, Jamaica, South Korea, Malaysia, and the Eastern Caribbean. Bound by low wages, poor working conditions, government repression, and multinational corporate extortion, they are trapped rather than mobile, isolated rather than connected. And with their ceaseless toil they create the material means not only of the gods' freedom but of their own enslavement. For it is their hidden effort that enables the multinational corporation, making possible global sourcing, rapid funds transfer, credit checking, just-in-time production, round-the-clock surveillance and security, online communication, and, of course, ubiquitous advertising.

But there is another side to this story. Beneath their electronic enhancements and omnipotent corporate personas, the gods are actually mortals themselves, however deformed by greed and power. Caught up in the chaos of unforgiving market competition, tangled in the web of cyberspeed-up, they have become inextricably dependent upon their ultimately uncertain deals and unreliable devices. They are thus at least as insecure, as isolated, as desperate as their mortal cousins. In their pursuit of divinity, moreover, they have sacrificed their humanity, their capacity for compassion, community, and cooperation, their ability to trust and be trusted. In the war between the gods, therefore, the combatants all march alone, without reliable relief or refuge.

There is war between the mortals as well. In the face of fickle, flexible, and footloose employers who are constantly on the lookout for the world's cheapest and most compliant labor, they vie with one another for jobs and investment, forever upping their ante of sacramental offerings to their gods. They can call upon few electronic miracles to gain advantage, only their resourcefulness and resolve, their skills, sweat, and sacrifice. But if the mortals cannot readily avail themselves of the wonders of the information age, neither are they dependent upon them for their identity or survival. They have retained their humanity, however strained, and hence a vestige of community and a potential for concerted action.

Finally, there is war between the gods and the mortals, and the gods, it seems, have every advantage. Omnipotent, omnipresent, omniscient, the gods are everywhere. They know everything. But look again, in this and their own internecine conflict they are not only despoiling the earth and degrading its inhabitants, but destroying each other. The formidable electronic extension of their rapacious reach, moreover, rests invariably upon an insecure foundation, already collapsing under its burden of cost and complexity. The vaunted power of the gods is precarious, moreover, because it is inescapably divided and divisive and thus perpetually susceptible to insider defection and deceit. For all their wondrous means of communication, their bonds of artifice and convenience, the gods can never really get it together.

The mortals, in contrast, stand together without accessories, whatever their handicaps. Theirs is the more primitive perhaps but also the more durable dimension, for here human bonds rest not upon electronic media but upon the unmediated and immediate: the senses, sociability, and the solidarity of shared struggle. Here power emanates not from the glowing screens of information networks but from the grins and grit of common skills and culture, and the deepest wells of defiance. The mortals have not escaped the bounds of space and time. They know where the weak points are in the divine connections, the particulars of place. They know when the right time comes, the importance of a moment. And they know, above all, what the gods themselves so easily forget—that the gods are not really gods. When their systems go down, they go down too.

Acknowledgements

The labor that went into producing this book was shared by the communication and information workers of the world. Their toil is embodied within the innards of the computers, monitors, diskettes, printers, and photocopiers on which our manuscripts were processed, stored, transmitted, and copied; in the telephone equipment and faxes that carried our voice and written messages; in the printshops where publications were created for our edification; in the broadcast, film, and animation production centers where spoken and visual ideas were produced for our memory; in the airline databanks that made possible our travel to faraway places; and in the many other artifacts that make up this wired and wireless world. We organized this project to share ideas with students of communications, economics, geography, international relations, labor studies, politics, sociology, third world studies, urban studies, and with workers everywhere.

We also would like to acknowledge individuals who helped shape the ideas that went into the organization of this project. They include Jomo Kwame Sundaram, Vinny Mosco, Joel Myron, Antonio Nieva, Jagdish Parekh, Rajah Rasiah, Etta Rosales, and Lenny Siegel, the animation and electronics workers in Asia who helped us understand the new international division of labor, and, with special appreciation, Connie Ozawa and Roseanne Lent, who supported and assisted in various stages of the book and helped us think through our own research and writing. Our deepest affections go to our children, Daniel and Jacqueline Sussman, and Laura, Andrea, John V., Lisa, and Shahnon Lent. John A. Lent wishes to add his appreciation to his families of origin for giving him insights into labor exploitation (having known it first hand as railway maintenance workers and coal miners) and to his mother, Rose Marie, and his late father, John.

—Gerald Sussman and John A. Lent

1

Global Productions

Gerald Sussman and John A. Lent

The collected essays in this book discuss the global structures of communication, information, and media through which pass daily transmissions of voice, data, text, and visual images—from the perspective of the people who actually build them. The "information society" is based on a new international division of labor of men and women sharing a production platform but dispersed into segmented zones of industrial, semi-industrial, and Third World societies. This geographical integration of labor processes, especially in such high tech industries as electronics and communications, represents the latest stage in the accumulation and flow of capital on a global scale.

Assembly workers in Hong Kong who manufacture the telecommunications, computer, and consumer electronics components and the telephone operators and repair workers who help traffic the action on stock, commodities, and currency exchanges get no commissions or royalty shares. People who build film and television sets on Hollywood, European, and Canadian lots or for international co-productions receive no "Oscars," not even credits or honorable mentions. In East Asia, where cartoon images for U.S. animation studios are reproduced by the millions, there are no free trips to the premieres in the "City of Dreams." Data entry workers in the Caribbean who process the information for airlines, banks, and insurance companies do not get frequent flyer mileage, have no money market accounts, and probably do not even have health care coverage. Communication workers of the world are people without citizenship in the "Global Village," yet they are the ones who make it possible.

The new international division of labor (NIDL), based on technological and telecommunications convergence, began to overlay the world economic system in the 1960s. Under the pressures of oligopolistic competition and falling rates of profit in the leading capitalist industrial states, the NIDL was organized by transnational corporations (TNCs) and state and supranational actors to expand the direction and speed up the motion of capital, overcoming heretofore spatial, temporal, and political limitations. With largely untapped human and physical resources available in the former colonies of the industrial powers and reduced communication and information costs, transnational capital coordinated its efforts in relocating investment to the Third World and to those peripheral areas in or near the West that were proximate to major markets (such as Ireland, Scotland, and the former Soviet bloc countries). The conquest of space with modern communication technology has opened areas once remote from the geographical reach of the core industrial powers. The global integration of production and a new international division of labor represents a fulfillment of the Columbian blueprint for world domination, now 500 years in passage.

Pursuing the logic of a singular, integrated, and telecommunications-linked world economic system, TNCs and their state allies have mobilized workers into a low-wage, segmented, and flexible global production force, made up of men and women who for the most part will never have the purchasing power to enjoy the goods and services created with their own labor. Of 2 million workers in heavily state-subsidized industrial parks (called export processing zones) in Third World and other less developed countries by the mid-1980s, two thirds were employed by TNC subsidiaries; a total of 7 million Third World workers were employed by TNCs overall. The region of Southeast Asia, for example, was producing 85% to 90% of all off-shore production of semiconductor assembly for Western consumption of computers and other consumer, military, and industrial electronics.

Export-oriented industrialization of these economies has also transformed the sexual composition of labor, starting in the 1960s, by bringing millions of young women into the industrial proletariat. Their participation rate in worldwide manufacturing doubled between 1970 and 1990, when they came to represent nearly half the workforce. Part of the explanation for NIDL, and which sets off this phase of transnational capitalism from earlier organizational forms of world economy, can be seen in the increased technology-based opportunities for accumulation on a global scale, propelled by the huge commercial spinoffs of decades of publicly funded Cold War military spending in transportation and communication (satellites, launch technology, space vehicles, ground stations, jet aircraft, fiber-optic cable, advanced electronics, computer hardware and software, digital switching and transmission systems, high-resolution Earth reconnaissance photography, etc.).

These technologies, many of them developed under government contracts with companies directly involved in mass media operations (General Electric, RCA, Hachette, Westinghouse) and put at the disposal of the transnational community to offset cyclical global economic downturns, have enormously expanded the value-added profit potentials of invested capital, while reducing dependence on location-specific skilled and unskilled labor pools. The 1980s' deregulation of freight hauling, airlines, telephony, computer, and media industries created wider cost-competitive and spatial control options, allowing AT&T and other telephone corporations, IBM and the computer manufacturing industry, film and TV production studios, satellite broadcasting, and other communication (and transportation) firms to branch out to new regional and technical markets. Electronics subsidiaries abroad provide over half of U.S. domestic consumption. Deregulation, a set of political rules designed to foster footloose capitalism, has encouraged the transfer of capital, labor, and consumer markets overseas (to countries that then export products back to the headquarter countries), forcing investment-hungry communities all over the world into common labor and consumption pools and unwanted competition based on lowered wage and benefit scales.

Modern telecommunications and electronic information networks make up the central nervous system of global capitalism. Central business districts of "world-class" cities require "intelligent" architecture that provides 24-hour, online connections to global commodities, financial, industrial, service, and information markets. When they can afford it, workers and middle-income households in the more affluent countries have access to post, telephone, and cable lines that provide opportunities for consumption-based leisure, and every credit transaction they initiate is computer-surveyed for recovery guarantees and consumer profiling. However, the information that is critical to citizenship, security, health, and well-being largely excludes those most in need by way of market-based "ability-to-pay" principles and educational redlining. In the meantime, income gaps between corporate CEOs and their employees continue to widen (in the U.S. averaging more than 200 to 1) as the tax burden on the wealthy diminishes year by year, which virtually assures workers that they will not be able to make ends meet or work their way out of debt, structural poverty, and marginalization.

Communication and information technology not only serves the accumulation goals of manufacturing and service industries but also provides the media of propaganda in defending the legitimacy and appeal of the transnational capitalist world order. Workers in mass media enterprises, including broadcasting, advertising, and public relations, sell not only products but images of the good life associated with those products. In the news media, people are sold images about the threats to and the defense of the moral foundations of society (e.g., police-thwarting criminals; the need to rein in the "demonic" behavior of Saddam Hussein, Fidel Castro, Manuel Noriega, and other infidels; the riotous instability of the Third World), along with spectacles of national and local politics and titillating accounts of daily life. Communication workers are involved at every nexus in the construction of these images, yet have little or no say over its underlying content. Electronics workers produce the equipment of the electronics media; telephone, data entry, and word processing employees handle the information throughput; film and TV workers make possible the means of visual production; and labor unions, when they exist, deliver the contracts and workers to their employers, sometimes with relatively decent wages and benefits for their members.

This book is an attempt to "deconstruct" the commodity fetish of communication and information technology by revealing the embedded labor components of their design, development, transfer, and usage. The contributors concentrated on histories of labor participation, structural relationships to the new international division of labor, conditions of work, the role of unions, and core issues within communication and information industries. Few textbooks pay attention to the political economic context in the study of communications, and fewer still acknowledge how mass media and information systems are functionally interlinked through the process of productive labor and how this process ultimately gives rise to the communications universe as we know it. *Global Productions* thus has an intended double meaning, referring both to the outward commodity form with which most people are familiar (TV programs, movies, VCRs, computers, etc.) and to the more hidden realm of the men and women who labor to make these commodities available for public and private use.

The chapters represent an extensive assortment of theory and praxis concerning the new international division of labor in communications, media, and information. They focus on the organization of labor in information-processing services, computer, industrial, and consumer electronics manufacturing and film, television, and animation production; the status of trade unions and workers in the restructured (following U.S.- and U.K.-style deregulation) Hollywood entertainment, British television, and Canadian telephone industries; and the efforts of the trade union movement itself to enter the information age. Using an array of research tools, including hosts of interviews, the chapters provide detailed case studies involving Barbados, Britain, Canada, China, Dominican Republic, Eastern Caribbean, Hong Kong, India, Ireland, Jamaica, Malaysia, South Korea, and the United States. The authors bring with them varied backgrounds, hailing from parts of Asia, the Caribbean, Europe, and North America: work in academic, trade union, and research and media professions: and share a common democratic concern for the well-being and dignity of working people everywhere.

Vincent Mosco gives an overview of the political economic approach to communication studies, defining this mode of analysis, charting its uses in communication research, and providing guidance on rethinking the philosophical roots of political economy and its use in understanding labor. He draws a map of political economy with three entry processes of commodification, spatialization, and structuration. Commodification of labor, according to Mosco, has shifted the balance of power in conceptualization from labor to "managerially controlled technological systems," as it disemploys media workers as part of its rationalizing design. Mosco argues that a consequence of spatialization is the development of global labor markets, enabling businesses to take advantage of differential wages, skill, and conditions of work. Structuration theory, he explains, thinks of society as an "ensemble of structuring actions initiated by social actors that shape and are shaped by class, gender, race, and social movement relations."

Mark Wilson is the first of the contributors to elaborate on the spatialization entry point, as he analyzes the global offshore labor force in information. His essay builds on the technical achievement of time - space compression, globalization of services, and the new international division of labor by observing the behavior of companies, in this case, those in the information processing services. He points out that the decreased costs of telecommunications has altered the rules about where companies can locate, while the globalization of services provides them with new labor supply advantages. Information processing employees in offshore centers servicing magazine subscription, insurance, and airline companies usually receive a small fraction of their U.S. counterparts' wages and benefits. Wilson describes some attributes of these so-called global back offices, discusses government incentives such as low taxes and exemptions from exchange controls and import duties that attract them, and concludes that offshore data processing has been important in the overall disaggregation of industrial, financial, and labor organization, separating and farming out basic employment to low-cost and nonmetropolitan environments.

Ewart Skinner brings the analysis of the new international division of labor in information production to a more specific level, that of the Caribbean offshore data processing/data entry industry. Based on his study of data work and his interviews in Barbados, Dominican Republic, Eastern Caribbean, and Jamaica, he develops a history of the Caribbean linkages to the data processing conduit. In the 1980s, governments of these states launched campaigns to attract U.S. data processing, promoting it as the technological coming of age for the Caribbean, rather than a deepening condition of dependency as some critics predicted.

To obtain the foreign work, according to Skinner, the governments made large investments in telecommunication technologies and highlighted the availability of a labor pool suffering from severe unemployment (as high as 30% in Dominican Republic) and underemployment and, thus, willing to accept these low paying jobs. He shows how foreign companies could afford to be choosy in such a labor climate, preferring Jamaicans and Barbadians because of their higher education and literacy rates, and women because of their presumed compliance and conscientiousness. Most of the jobs, Skinner finds, are closely monitored, fast-paced, tedious, repetitive, routinized, unsafe, and unhealthy.

Work and labor in the world's "number one address in high technology," Silicon Valley, is the focus of Lenny Siegel's study. His major contention is that the U.S. union movement has been misled into believing that all is well in Silicon Valley on the basis of what it sees as the high living standards of white collar workers; however, as he points out, the reality is different for most service employees. Arguing that the region's business culture is "less about new age meritocracy and more about race and gender discrimination," he shows how the Valley hires a large number of poorly paid workers, most of whom are women, immigrants, and non-white minorities.

Siegel finds that lower income employees have added disadvantages brought about by their domicile. Forced to live farther from their jobs, where housing is cheaper, compared to their white-collar counterparts, they are required to spend much of their day commuting. At the same time, their communities, with weaker tax bases, do not receive close to the same positive benefits, such as good public schooling, generated in better suburbs by the companies in which they work. As Silicon Valley companies have expanded, according to Siegel, they have relocated segments of their production to other places where low-cost labor is available, or they have farmed out "dirty work" to subcontractors, thus not only getting the cheaper labor, but also avoiding the social responsibility for environmental pollution and workplace health hazards.

Gerald Sussman's case study examines the workforce of computer, industrial, and consumer electronics factories in Malaysia. Like Skinner, he describes how the government, in the Malaysian case, enticed transnational corporations (TNCs) by selling the "dexterous hands of the oriental female" and, less publicly, by emphasizing her lower propensity to organize or rebel against the highly stressful and dangerous shopfloor conditions. Sussman makes the case that although low wages are important for investment decisions made by TNCs, the overriding factor is state behavior—promoting stable government, industrial peace, and anti-union policies.

Employing class, gender, and ethnic politics, the Mahathir government has assured such workplace stability by opposing a national electronics union, resorting to arrest and torture of labor organizers, playing off Malay against Chinese citizens, and keeping a no-blink vigil so that no aspect of the social or political fabric gets out of control. The benefits from the formation of the export processing zones in Malaysia have eluded most Malaysians, as very little technology transfer or new domestic value-added enterprise have accrued. Laborers (nearly three fourths of whom are female) are compensated at 10% to 12% of their Western counterparts, yet they are expected to carry a 50% heavier workload. Sussman explains that increased automation has perpetuated dependency relations as it has intensified the work pressure and created few transferable skills beyond the plants.

Lai Si Tsui-Auch focuses on another region of Asia—Hong Kong and the Shenzhen Special Economic Zone of China—to study subcontracting and labor in the production of information/communication technology (I/CT). Her research is based on data collected during 1994 interviews with more than 40 manufacturers. Historically, Hong Kong was second after Japan among Asian states involved in the production of IC/T; by the late 1960s, the British colony became the principal assembler of semiconductors for the U.S. market. During the 1980s, 80% of Hong Kong electronics manufacturers kept competitive by moving assembly processes north to Shenzhen and the Pearl River Delta regions on China's mainland.

Predictably, as Tsui-Auch notes, the number of Hong Kong factories dropped dramatically, as did the number of workers, with relocation to China and with increased automation. The worker conditions she identifies resemble those elsewhere—workers are asked to put in two to four hours daily overtime, are pressured toward higher levels of quality and efficiency, and are subjected to an intruding surveillance system, unsafe work conditions, low wages, inhuman living conditions, and physical abuse. Much like Chinese male laborers in the 19th-century United States, they are discouraged from marrying and establishing families in the Special Economic Zone and are penalized with higher school fees for their children if they do.

Janet Wasko tells us that although the global markets for Hollywood's products have grown by leaps and bounds, the picture for trade unions and workers in the entertainment industry is gloomy. From the beginning, Hollywood labor has had a difficult time, first, getting recognized in the 1930s and, second, surviving ideological assaults and blacklisting of the 1940s and 1950s. Wasko analyzes the Hollywood labor situation along the lines of various unions and guilds and focuses on issues of non-union and non-Hollywood production, corporate diversification, and the internationalization of the entertainment business. One disturbing finding in her research is that Hollywood unions are on the decline, and that less than a third of the films released in the United States are produced with union labor.

Margaret Thatcher's policies regarding unions come under severe criticism in James Cornford and Kevin Robins's treatment of the changing labor force in British television. They argue that British broadcasting had been sustained by a stable set of compromises until the implementation of Thatcher's strategy of picking off organized labor groups, one by one. Turning her attention to communications after unions in the manufacturing and mining sectors were destroyed, Thatcher received help from media proprietors such as Bruce Gyngell, Rupert Murdoch, and Eddy Shah, who mounted management offensives that included moving newspaper plants out of Fleet Street, setting up their own distribution systems, replacing printers with workers from the Electricians' Union, and locking out striking television workers. The scenario was completed when broadcasting was restructured with a government-supported emphasis on independent production: thus, a serious reduction in permanent employment resulted. In the process, Cornford and Robins contend, television has been killing its most important assets - the skills of its workers. Ironically, with the resultant drop in production values, the government's aim of making British broadcasting more export oriented has been a failure.

Using empirical data and interviews, Manjunath Pendakur sheds light on the implications of the changes in relations between capital and labor in film and television production. He hones in on the relocation of U.S. film and television production to British Columbia, Canada. The reason for such investment in British Columbia is simple - the allure of getting more for less by way of flexible government policies and labor pools.

However, Pendakur says there are other reasons why American producers prefer shooting in the Canadian province, which include the weakened Canadian-U.S. dollar exchange, the diverse beauty of British Columbia, the availability of highly skilled labor, and relief from the high crime rate identified with Los Angeles. British Columbia works hard to entice foreign filmmakers; both federal and provincial governments advertise attractive tax write-off incentives. What is somewhat surprising, however, is the malleability of the labor unions. Pendakur reports from talks with British Columbia union officials that they had collaborated with the British Columbia Film Commission, accepting major concessions to attract foreign producers. One of the key concessions is acceptance of flexible contracts, the underlying belief being that creating employment is more important than fighting for better wages and working conditions. This type of thinking, Pendakur concludes, has resulted from the weakened position of Canadian and international unions, which has placed them in a struggle for survival and left them unprepared to negotiate with the powerful and expanding entertainment industry.

Usually ignored in discussions of labor in communications, media, and information production is the animation sector. Yet, as John A. Lent points out, animators have been some of the most exploited creators in the arts and media for many years, working without credit for such studio heads as Disney, Schlesinger, or Quimby. Serious disagreements with management, going back before World War II, sometimes led to strikes on issues of wages and working conditions.

Since the 1960s, and increasingly after the mid-1970s, West Coast studios have pursued a new international division of labor strategy in subcontracting the bulk of their work to offshore plants, first in Japan, and later in Taiwan, South Korea, and Australia. Other countries such as Canada, France, and Britain followed Hollywood's lead, attracted, as Lent explains, to cheaper labor, good artistic skills, and mobility incentives. When labor costs in Taiwan escalated, studios there set up branches in and subcontracted work to China, the Philippines, and Thailand. The U.S. and European studios have not been the only benefactors of this arrangement, according to what some Asian animators told Lent. According to them, offshore animation production has transferred skills locally, among individuals who, in some cases, have set up domestic animation operations of their own. It has also brought in needed foreign capital and given employment to many young people.

Sid Shniad discusses worker experiences at the British Columbia Telephone Company and the impacts of corporate-driven restructuring of Canadian telecommunications. As have others in this book, Shniad approaches the subject first from an historical perspective. He has high praise for the Canadian telephone system before it initiated U.S.-style deregulation, showing how long-distance profits were used to subsidize local and residential services. In his estimation, Canada provided technically sophisticated services at some of the most affordable rates in the world.

All this changed, however, in the shift to deregulation, increased competition, and globalization, all of which were highly beneficial to large-scale business accounts and not the ordinary telephone users of Canada. Workers told Shniad how telephone companies have reduced costs with tens of thousands of layoffs, at the same time catering to stockholder demands for the introduction of newer technologies and services. Those who survived the layoffs said they now work in a tightly monitored, pressurized environment for lower wages, with no guarantees regarding their health insurance, pension plan, or job security.

Finally, Dave Spooner tells us how the trade union movement is joining the "information age," using the tools of information technology and telecommunications to conduct its own business. He provides telematics strategies for unions in organizing situations ranging from the internationally run "geocentric" to the more locally managed "polycentric" firms and discusses bargaining opportunities in new but integrated "greenfield" subsidiaries as compared to those loosely attached by merger or acquisition. Spooner envisions trade union telematics developing a "complex and shifting web of communication networks and information sources" linking national unions, workplaces, nongovernment organizations, and other popular associations. He is also realistic about the constraints, such as costs, lack of network access in many countries, inadequate training, lack of political awareness, language differences, and the reluctance of national trade unions to cede authority to international trade union structures. Spooner further suggests elements needed in a trade union telematics strategy that can facilitate larger spheres of labor participation.

Together, the chapters in *Global Productions* open the analysis of communication and information technology to a wider set of social issues, regardless of what analysis - Marxian, Rawlsian, or other - one prefers. Telecommunications and "information society" restructuring offer transnational corporations new opportunities for integrating and "third worldizing" labor forces in the leading and emerging industrial countries, as states surrender their relatively independent status vis-à-vis the growing spatial and temporal demands of world capital. State leaders in Third World countries are not helpless, however, and many make comfortable personal accommodations to these new global economic arrangements, as they have in the past, whereas others display a degree of national resolve. Workers, too, are not helpless as they find new means for internationalizing their potential collective strength and in appropriating communication and information technologies as their own instruments in their struggle for democratic political participation and a decent and sustainable quality of life.

2

Political Economy, Communication, and Labor*

Vincent Mosco

This chapter provides an overview of the political economy approach to communication studies, which has informed most research on labor in the communication industries and on the role of communication in the international division of labor.

The chapter is divided into two sections, the first of which defines political economy and charts its use in communication research. The second provides guidance for rethinking and renewing the philosophical roots and substantive terrain of the approach and suggests how this might be used to understand labor. The time is ripe for such a rethinking of political economy because transformations in the world

*The author acknowledges a grant from the Canadian Social Sciences and Humanities Research Council for research on the political economy of communication. This chapter summarizes and applies to labor arguments contained in the author's book, *The Political Economy of Communication: Rethinking and Renewal*, London: Sage, 1996.

political economy and in intellectual life have raised fundamental challenges to the approach. The global political economy has been marked by the demise of communist party rule in the former Soviet bloc, the continued stagnation in capitalist societies, the breakup of what unity once existed in the Third World, and the rise of social movements, particularly feminism and environmentalism, that cut across traditional political economic categories such as social class.

Among the numerous intellectual challenges to political economy, two stand out for their significance in examining labor in the communications industries. On one side, cultural studies challenges the institutional ground of political economy and questions the centrality of labor as a historical force. On the other, an approach variously called policy science, public choice theory, rational expectations, and "positive" political economy applies neoclassical economic theory to social behavior and thereby reduces labor to one, less than central, element in a pluralist marketplace of individual choice.

The first section begins by defining the political economy approach, identifying its fundamental characteristics and mapping major schools of thought. From there, it proceeds to examine how communication scholars have drawn on the theoretical framework to carry out research on mass media and information technologies. This section highlights divisions that distinguish research approaches in North America, Europe, and the Third World. The second section begins the process of rethinking the political economy of communication by proposing how to address its philosophical foundations. Specifically, it calls for an epistemology based on a realist, nonessentialist, inclusive, and critical approach to knowledge and an ontology that foregrounds social change, social process, and social relations over the traditional tendency in political economy to start from structures and institutions. Putting this agenda into practice, the chapter identifies three processes that constitute the central entry points for political economy research—commodification, spatialization, and structuration—and suggests how they can be used to comprehend labor and communication, including the international division of labor. The chapter concludes by describing how this renewed political economy of communication responds to challenges on its borders from cultural studies and policy science.

WHAT IS POLITICAL ECONOMY?

Two definitions of political economy capture the wide range of specific and general approaches to the discipline that social theory presents. In the narrow sense, *political economy* is the study of the social relations, particularly the power relations, that mutually constitute the production, distribution, and consumption of resources, including communications resources. This formulation calls attention to the institutional circuit of communications products that link, for example, a chain of primary producers to wholesalers, retailers, and consumers, whose purchases, rentals, and attention are fed back into new processes of production. However, there is sufficient ambiguity about what constitutes a producer, distributor, or consumer that one needs to be cautious in using it.

A more general and ambitious definition of *political economy* is the study of control and survival in social life. Control refers specifically to the internal organization of people and the process of adapting to change. Survival means how people produce what they need for social reproduction and continuity. In this reading, control processes are broadly political in that they constitute the social organization of relationships within a community and survival processes are fundamentally economic because they concern processes of production and reproduction. The strength of this definition is that it gives political economy the breadth to encompass at least all human activity and, arguably, all organic processes, a tendency reflected in environmental, ecological, and biodiversity studies (Benton, 1989).[1] Its principal drawback is that it can lead one to overlook what distinguishes human political economy from general processes of survival and control. This includes, particularly, the unique goal-oriented power of consciousness, which is literally aware of its own awareness.

Another way to describe political economy is to broaden its meaning beyond what is typically considered in definitions by focusing on a set of central qualities that characterize the approach. Drawing particularly on the work of Golding and Murdock (1991), this section focuses on four ideas—social transformation, the social totality, moral philosophy, and praxis—which different schools of political economic thought tend to share.

Political economy has consistently placed in the foreground the goal of understanding social change and historical transformation. For classical political economists such as Smith, Ricardo, and Mill, this meant comprehending the great capitalist revolution, the vast social upheaval that would transform societies based primarily on agricultural labor into commercial, manufacturing, and, eventually, industrial societies. For Marx, it meant examining the dynamic forces within capitalism and between it and other forms of political economic organization in order to understand the processes of social change that would, ultimately, transform capitalism into socialism.

---

[1]This definition grows out of a suggestion that Dallas Smythe offered in a series of discussions in December 1991.

Orthodox economics, which began to coalesce against political economy in the late 19th century, tended to set aside this concern for the dynamics of history and social change in order to turn political economy into the science of economics, whose law-like statements were best constructed to fit static rather than dynamic social conditions. Contemporary political economists, occupying various heterodox positions distinct from what has become economic orthodoxy, continue the tradition to take up social transformation. However, humbled by the failure of the 20th century to live up to the utopian visions of the Enlightenment and its offspring, they tend to focus attention more modestly around the axes formed by various "post" formulations, particularly postfordism, postindustrialism, and postmodernism.

Political economy is also characterized by an interest in examining the social whole or the totality of social relations that constitute the economic, political, social, and cultural fields. Academic orthodoxy tends to compartmentalize these into distinct disciplines, each with its own rules of entry, border controls, and systems of overall surveillance. Currently fashionable poststructuralist thought, although laudable in its attack on this and other orthodoxies, also tends to deny the existence of social, or even of discursive, totalities that argue for principles of order across the range of discrete social practices. From the time of Adam Smith, whose interest in understanding social life was not constrained by disciplinary boundaries, through Marx, and on to contemporary institutional, conservative, and neo-marxian theorists, political economy has consistently aimed to build on the unity of the political and the economic by accounting for their mutual constitution and for their relationship to wider social and symbolic spheres of activity.

Political economy is also noted for its commitment to moral philosophy, understood as both an interest in the values that help to constitute social behavior and, normatively, in those moral principles that ought to guide efforts to change it. For Adam Smith, as evidenced in his *Theory of Moral Sentiments* (1759/1976), a book he favored more than the popular *Wealth of Nations* (1776/1937), this meant understanding values such as acquisitiveness and individual freedom, which were contributing to the rise of commercial capitalism. However, for Marx (1973, 1976), moral philosophy meant the ongoing struggle between the drive to realize self- and social value in human labor and the drive to reduce labor to a marketable commodity. Contemporary political economy tends to moral philosophical standpoints that foreground the extension of democracy beyond the political realm, where it is variously legitimated in formal, legal instruments to encompass the economic, social, and cultural domains that tend to be shaped by the requirements of capital.

Following from this view, social praxis, or the fundamental unity of theory and practice, also occupies a central place in political economy. Specifically, opposed to orthodox positions, which separate, at least formally, the sphere of research from social intervention, political economy consistently viewed intellectual life as a conduit to social transformation and social intervention as a conduit to knowledge. This tendency has a long history, dating from centuries-old practices of providing advice and counsel to power. Although they differ fundamentally on what should characterize intervention, from Malthus, who supported open sewers as a form of population control, to Marx, who called on labor to realize itself in revolution, political economists are united in the view that the division between research and action is artificial and must be overturned.

The political economy approach is also distinguished by the many schools of thought that guarantee significant variety of viewpoints and vigorous internal debate. Arguably the most important divide emerged in 19th-century responses to the classical political economy of Smith and his followers. One set aimed to build on the classical emphasis on the individual as the primary unit of analysis and the market as the principal structure. According to this view, the central *process* is the individual decision to register wants or demands in the marketplace. Over time, this response progressively eliminated the classical concerns for history, the social totality, moral philosophy, and praxis in order to transform political economy into the science of economics founded on empirical investigation of marketplace behavior conceptualized in the language of mathematics. This approach, broadly understood as neoclassical economics or simply, in recognition of its hegemonic position as the orthodoxy, economics, reduces labor to one among the factors of production, which, along with land and capital, is valued solely for its productivity or the ability to enhance the market value of a final product (Jevons, 1965; Marshall, 1890/1961). Although tensions exist within economic orthodoxy, including, from time to time, challenges from more institutionally grounded work (Keynes, 1964; Schumpeter, 1942), the mainstream position is notable for its disciplinary authority.

A second set of responses opposed this tendency by retaining the classical concern for history, the social whole, moral philosophy, and praxis, even if that meant giving up the goal of creating a positive science. This set constitutes the wide variety of approaches to political economy. A first wave was led by conservative followers of Edmund Burke, who replaced marketplace individualism with the collective authority of tradition (Carlyle, 1984); by Utopian Socialists, who accepted the classical faith in social intervention but urged putting community ahead of the market (Owen, 1851); and by marxian thought, which returned labor to the center of political economy. According to the latter, *Homo Faber*, or "man the maker," defined human species-being, specifically the unique integration of conception and execution that separated, in Marx's example, the architect from the bee. Subsequent formulations built on these perspectives, leaving us with a wide range of contemporary formulations.

Although orthodox economics occupies the center and center-right of the intellectual spectrum, a conservative political economy thrives in the work of people like George J. Stigler (1988), James M. Buchanan (Brennan & Buchanan, 1985), and Ronald Coase (1991; Coase & Barrett, 1968)), recent recipients of the Nobel prize in economics, who would apply the categories of neoclassical economics to all social behavior with the aim of expanding individual freedom. Institutional political economy occupies a slightly left of center view, arguing, for example, in the work of Galbraith (1985), who drew principally on Veblen (1932, 1899/1934), that institutional and technological constraints shape markets to the advantage of those corporations and governments with the power to control them.

Among their accomplishments, institutionalists produced economic histories of labor and trade unions that challenged the narrow, individualistic conception of neoclassical economists. Neo-marxian approaches, including the French Regulation School (Lipietz, 1988), world systems theory (Wallerstein, 1979), and others engaged in the debate over Fordism (Foster, 1988), continue to place labor at the center of analysis and are principally responsible for debates on the relationship between monopoly capitalism, deskilling, and the growth of an international division of labor. Finally, social movements have spawned their own schools of political economy, primarily feminist political economy, which addresses the persistence of patriarchy and the unrecognized economic value of household labor (Waring, 1988), and environmental political economy, which concentrates on the links between social behavior and the wider organic environment (Benton, 1989).

THE POLITICAL ECONOMY OF COMMUNICATION

Communication studies has drawn on the various schools of political economic analysis. This section concentrates on research situated in the institutional and neo-marxian approaches because these have paid the closest attention to labor in the communication industries and to the impact of communication on contemporary transformations in labor. Although both neoclassical economists (Owen & Wildman, 1992) and conservative political economists (Coase & Barrett, 1968) have theorized the communication industries, they have not situated labor within these analyses.

At this stage in its development, it is useful to map the political economy of communication from the perspective of regional emphases. Although there are important exceptions and cross-currents, North American, European, and Third World approaches differ enough to receive distinctive treatment. Moreover, the political economy approach to communication is not sufficiently developed theoretically to be explained in a single analytical map.

North American research has been extensively influenced by the contributions of two founding figures, Dallas Smythe and Herbert Schiller. Smythe taught the first course in the political economy of communication at the University of Illinois and is the first of four generations of scholars linked together in this research tradition.[2] Schiller has similarly influenced several generations of subsequent political economists. Their approach to communication studies draws on both the institutional and marxian traditions. However, they have been less interested than, for example, European scholars, in providing a theoretical account of communication. Rather, their work and, through their influence, a great deal of the research in this region has been driven more explicitly by a sense of injustice that the communication industry has become an integral part of a wider corporate order that is both exploitative and undemocratic. Although they have been concerned with the impact within their respective national bases, both have led a research program that charts the growth in power and influence of transnational media companies throughout the world (Smythe, 1981; Schiller, 1989, 1969/1992).

Partly owing to their influence, North American research has produced a large literature on industry and social class-specific manifestations of transnational corporate and state power, distinguished by its concern to participate in ongoing struggles, including those of labor to change the dominant media and to create alternatives (Douglas, 1986; Mosco & Wasko, 1983; Wasko and Mosco, 1992). A major objective of this work is to advance public interest concerns before government regulatory and policy organs. This includes support for those movements that have taken an active role before international fora, such as the United Nations, in defense of a new, more equitable international economic, information, and communication order (Roach, 1993).

---

[2]Smythe's student Thomas Guback taught the political economy of film at the University of Illinois. Janet Wasko, a student of Guback, also works in this area at the University of Oregon, while a student of hers, Jack Banks, does research on the political economy of music at the University of Hartford.

European research is less clearly linked to specific founding figures and, although it is also connected to movements for social change, particularly defense of public service media systems, the leading work in this region has been more concerned with integrating communication research within various neo-marxian theoretical traditions. Of the two principal directions this research has taken, one, most prominent in the work of Garnham (1990) and Golding and Murdock (1991; Murdock & Golding, 1979), has emphasized class power. Building on the Frankfurt School tradition, as well as on the work of Raymond Williams, it documents the integration of communication institutions, mainly business, and state policy authorities, within the wider capitalist economy, and the resistance of subaltern classes and movements reflected mainly in opposition to neo-conservative state practices promoting liberalization, commercialization, and privatization of the communication industries. A second stream of research foregrounds class struggle and is most prominent in the work of Armand Mattelart (Mattelart & Mattelart, 1986/1992; Mattelart & Siegelaub, 1983). Mattelart has drawn from a range of traditions including dependency theory, Western marxism, and the worldwide experience of national liberation movements to understand communication as one among the principal sources of resistance to power.

Among the class power work directed specifically at an understanding of labor from this perspective, Miège (1989) offers an assessment of different labor processes that tend to cohere with different forms of media production within the overall logic of capitalist social relations. He suggests that there is a connection between the type of media product, the structure of corporate control, and the nature of the labor process. Media hardware such as television receivers and recorders is characterized by a simple process of production and little intervention of creative or artistic workers. According to this view, these media products lend themselves to industrial concentration and a detailed labor process, including an international division of labor that takes advantage of low-wage areas with predominantly nonunionized workers subjected to a regime of authoritarian control. At the opposite end of the spectrum, a second type of product, art prints and what he calls "realisations audio-visuelle," is produced almost solely with artisanal or craft labor, is not easily reproducible, and requires relatively low infusions of capital. This supports a sector dominated by small businesses and enables widespread producer or labor control. Miège identifies a final product type as a principal site of struggle and conflict because it is both easily reproducible and requires some degree of artistic contribution. This sector contains growing, but far from complete, monopoly control and a wide mix of labor that makes for tensions and conflicts within, as well as between, capital and labor. Research on labor and class struggle has been prominent in the work of Peter Waterman (1990, 1992), who has documented labor and trade union use of the mass media and new communication technologies to promote democracy and internationalism.

Third World research on the political economy of communication has covered a wide area of interests, although a major stream has grown in opposition to modernization theory, an approach that originated in Western, particularly U.S., attempts to incorporate communication into an explanatory paradigm congenial to mainstream intellectual and political interests. Modernization theory held that the media were resources that, along with urbanization, education, and other social forces, would mutually stimulate progressive economic, social, and cultural modernization. As a result, media growth was viewed as an index of development. Drawing variously on dependency, world systems, and other streams of international neo-marxian political economy, Third World political economists challenged the fundamental premises of the model, particularly its technological determinism and the omission of practically any interest in the power relations that shape the terms of relationships between First and Third World nations and the multi-layered class relations between and within them (Boafo, 1991; Cardoso & Faletto, 1979; Roncagliolo, 1986; Tang & Chan, 1990).

The failure of development schemes incorporating media investment sent modernization theorists in search of revised models that have tended to incorporate telecommunications and new computer technologies into the mix (Jussawalla, 1986). Political economists have responded mainly by addressing the power of these new technologies to integrate a global division of labor. A first wave of research saw the division largely in territorial terms: unskilled labor concentrated in the poorest nations, semi-skilled and more complex assembly labor in semi-peripheral societies, and research, development, and strategic planning limited to First World corporate headquarters to which the bulk of profit would flow. More recent research acknowledges that class divisions cut across territorial lines and maintains that what is central to the evolving international division of labor is the growth in flexibility for firms that control the range of technologies that overcome traditional time and space constraints (Harvey, 1989; Morris-Suzuki, 1986; Sivanandan, 1990).

RETHINKING POLITICAL ECONOMY

Most assessments of political economy, including its application to communication research, acknowledge its contribution to intellectual life and political struggles. Nevertheless, these also raise concerns about the need to rethink and renew the discipline in light of recent upheavals. This section responds to this general ferment by suggesting starting points for rethinking political economy that can guide research in communication, including the relationship to labor.

The philosophical foundations of a political economy approach to communication provide an important starting point. Drawing on recent critical literature that reflects the state of the field, I advance basic epistemological and ontological principles (Gandy, 1992; Golding & Murdock, 1991). An epistemology is a way of knowing or an approach to understanding a field of knowledge. The political economy of communication needs to be grounded in a realist, inclusive, constitutive, and critical epistemology.

Political economy is realist in that it recognizes the reality of both concepts and social practices, thereby eschewing idiographic and nomothetic approaches, currently fashionable in poststructural thought, which argue respectively for the reality of discourse only or reject the reality premises of both concepts and practices. Following from this, political economy is inclusive in that it rejects essentialism, which would reduce all social practices to a single political economic explanation in favor of an approach that views concepts as entry points into a social field (Resnick & Wolff, 1987). The choice of certain concepts and theories over others means that one gives priority to these as useful means of explanation. They are not assertions of the one best, or only, way to understand social practices.

Additionally, the epistemology is constitutive in that it recognizes the limits of causal determination, including the assumption that units of social analysis interact as fully formed wholes and in a linear fashion. Rather, it approaches the social as a set of mutually constitutive processes, acting on one another in various stages of formation and with a direction and impact that can only be comprehended in specific research.

Finally, the approach is critical in that knowledge, the mutual constitution of theory and practice, is viewed as the product of an ongoing set of comparisons to other bodies of knowledge and to a set of normative considerations that guide social praxis. For example, political economy is critical in that it regularly situates the knowledge acquired in research against alternative bodies of knowledge in, for example, neoclassical economics, pluralist political science, and cultural studies. Furthermore, it measures political economic knowledge against the values of social democracy, including public participation and equality, that guide social praxis.

Connected to this epistemology, this theory of knowledge, is an ontology or a theory of being. Political economy needs an ontology that foregrounds social change, social process, and social relations against the tendency in research, particularly in political economy, to concentrate on structures and institutions. This means that research starts from the view that social change is ubiquitous, that structures and institutions are in the process of constant change, and that it is therefore more useful to develop entry points that characterize processes rather than to name institutions.

Guided by this principle, we can consider a substantive map of political economy with three entry processes, starting with commodification, the process of transforming use to exchange value. From this, we can move on to spatialization, the transformation of space with time or the process of institutional extension, and finally to structuration, the process of constituting structures with social agency. Foregrounding social change with these processes does not replace structures and institutions, something that would substitute one form of essentialism for another. Rather, these are entry points that constitute a substantive theory of political economy, one preferred choice among a range of possible means of understanding the social field. This section takes up these substantive entry points, using them to suggest the boundaries of a political economic analysis and, more specifically, to understand labor and the international division of labor in communication research.

Commodification has long been understood as the process of taking goods and services that are valued for their use, for example, food to satisfy hunger, and transforming them into commodities that are valued for what they can bring in the marketplace. The process of commodification holds a dual significance for communication research. First, communication practices and technologies contribute to the general commodification process throughout society. For example, the introduction of computer communication gives companies greater control over the entire circuit of production, distribution, and exchange, permitting retailers to monitor sales, inventory levels, and workers with ever-improving precision, and enabling them to produce and ship only what they know is likely to sell quickly, thereby reducing inventory requirements and unnecessary merchandise. Second, commodification is an entry point to understanding specific communication institutions and practices. For example, the general, worldwide expansion of commodification in the 1980s, responding in part to global declines in economic growth, led to the increased commercialization of programming, the privatization of once public media and telecommunications institutions, and the elimination of workers.

The political economy of communication has been notable for its emphasis on describing and examining the significance of those structural forms responsible for the production, distribution, and exchange of communication commodities and for the regulation of the communication marketplace. Although it has not neglected the commodity itself and the process of commodification, the tendency has been to foreground corporate and state structures and institutions. When it has treated the commodity, political economy has tended to concentrate on media content, to a lesser extent on media audiences, and paid surprisingly little attention to the labor process.

The emphasis on media structures and content is understandable in light of the importance of global media companies and the growth in the value of media content. Tightly integrated transnationals, such as Time Warner, create media products with a multiplier effect embodied, for example, in the tiered release, which starts with a theatrical film exhibited in U.S. and foreign theaters, followed in six months or so by a video, shortly thereafter released on pay-per view, pay cable, and finally perhaps aired on broadcast television.

Political economy has paid some attention to audiences, particularly in explaining how advertisers "purchase" the quality and quantity of audiences that newspapers, magazines, radio, or television programs can deliver. This has generated a vigorous debate about whether audiences, in fact, constitute a form of "labor," in effect involving the selling of their labor power (their attention) in exchange for media content (Murdock, 1978; Smythe, 1997). The debate has been useful because it broadened the discussion beyond content, thrusting advertisers, as the general representatives of capital and not merely a media fraction, into the core of communication research. However, the debate disinclined traditional political economists from media research because it raised questions about the labor theory of value and took attention away from the sustained examination of commodification and labor. The latter is historically understood to comprise those who have progressively lost control over the means of production and are consequently left with only their labor power, which they sell for a wage (Lebowitz, 1986).

Until the 1970s, political economic research, with the possible exception of the institutionalist school, in the interest of examining the commodification of goods, continued to neglect the labor commodity and the process that takes place at the point of production. Braverman's (1974) work gave rise to an intellectual drive to end this marginal status by directly confronting the transformation of the labor process in capitalism. According to him, labor is constituted out of the unity of conception (the power to envision, imagine, and design work) and execution (the power to carry it out). In the process of commodification, capital acts to separate conception from execution, skill from the raw ability to carry out a task; to concentrate conceptual power in a managerial class fraction that is either a part of capital or that represents its interests; and to reconstitute the labor process with this new distribution of skill and power at the point of production. In the extreme, and with considerable labor resistance, this involved the application of so-called scientific management practices, pioneered by Frederick Winslow Taylor.

Braverman documented the process of labor transformation in the rise of large-scale industry, but he is particularly recognized for producing one of the first sustained examinations demonstrating the extension of this process into the service and information sectors. His work gave rise to an enormous body of empirical research and theoretical debate, the latter focusing mostly on the need to address the contested nature of the process, the active agency of workers, and the trade union movement (Burawoy, 1979; Edwards, 1979), as well as on how the transformation of the labor process was experienced differently by industry, occupation, class, gender, and race (Berberoglu, 1993).

The political economy literature shows some evidence of this works' influence, particularly in research on the introduction of new communication and information technologies and on the transformation of work, including patterns of employment and the changing nature of labor, in the media and telecommunications industry.

Decrying the absence of a labor perspective in journalism history, Hardt (1990) aims to fuse what is essentially a political economic perspective with a cultural history of the newsroom that focuses on the introduction of new technologies deployed to carry out the processes Braverman described. This extends the pioneering work of political economists working outside communication studies, who have studied the labor process in the newsroom (Zimbalist, 1979).

More recent work that, inter alia, addresses the commodification of labor in the newsroom looks to the application of new technologies to reduce employment in the industry and to restructure the work of editors by implementing electronic page layout or pagination and, to a lesser extent, of reporters with electronic news gathering (Russial, 1989). These are specific applications of the labor process view that points to the use of communication and information technologies that shift the balance of power in conceptual activity from, in this case, professional newsworkers with some control over their means of communication to managerially controlled technological systems.

Similar work has begun to address the transformation of the labor process in film (Nielsen, 1990), broadcasting (Wasko, 1983), telecommunications (Mosco & Zureik, 1987), and the information industries (Kraft & Dubnoff, 1986). As noted earlier, Miège (1989) offers a variation on this analysis, amounting to an effort to bridge political economy and organizational communication research (cf. Fishman, 1980) by suggesting that there is a connection between the type of media product, the structure of corporate control, and the nature of the labor process.

The second substantive entry point is spatialization, or the process of overcoming the constraints of space and time in social life. Classical theorists such as Smith and Ricardo, partly in response to their 18th-century physiocratic predecessors (who saw economic value only in land), found it necessary to devote considerable attention to the problems of how to value space, and their development of a labor theory of value was bound up with the problem of how to define and measure labor time. Marx came closer to spatialization when, in the *Grundrisse* (1973), he noted that capitalism "annihilates space with time." By this, he meant that capital makes use of the means of transportation and communication to diminish the time it takes to move goods, people, and messages over space. Recent theorists (Lash & Urry, 1987) modify this view by suggesting that, rather than annihilate space, capital transforms it. They remind us that people, products, and messages have to be located somewhere, and it is this somewhere that is undergoing significant transformation, evidenced in, for example, upheavals in the international division of labor.

Spatialization is similar to concepts offered by geographers and sociologists to address structural changes brought about by shifting uses of space and time. Giddens (1990) refers to the centrality of time-space distanciation in order to examine the decline in human dependency on space and time. He suggests that this process expands the availability of time and space as resources for those who can make use of them. Harvey (1989) identifies time-space compression to suggest how the effective map of the world is shrinking, again for those who can take advantage of it. Castells (1989) calls our attention to the declining importance of physical space, the space of places, and the rising significance of the space of flows to suggest that the world map is being redrawn according to boundaries established by flows of people, goods, and services, creating what Massey (1992) refers to as a transformed "power geometry."

Communication is central to spatialization because communication and information processes and technologies promote flexibility and control throughout business and particularly within the communication and information businesses themselves. Spatialization encompasses the much-abused term *globalization*, which perhaps best refers to the worldwide restructuring of industries and firms. Restructuring at the industry level is exemplified by the development of integrated markets based on digital technologies and, at the firm level, by the growth of the flexible or "virtual" company, which makes use of communication and information systems to continuously change structure, product line, marketing, and relationships to other companies, suppliers, its own workforce, and customers.

There is little disagreement with the conclusion that by accelerating the introduction of communication and information technologies, many firms, whether in manufacturing or services, have been able to change the production process and the relationship between capital and labor. In general, this is viewed as a shift away from a Fordist system, named after that pioneer in assembly line manufacturing Henry Ford, which concentrated on mass production of homogenous goods by workers organized in single task hierarchies. Replacing Fordism are new systems of production in small, customized batches and a labor process featuring multiple tasks, continuous training, and the elimination of rigid job demarcations. Aptly named flexible accumulation, these systems allow business to take advantage of technologies that overcome space and time constraints. Business can now draw on capital, material, and labor practically wherever and whenever it needs to and to sell products to market segments around the world.

Analysts now debate the significance of this transformation. Does it signal a fundamental change in capitalism or a deepening and extension of centuries-old practices? The answer to this question is of particular significance for the communication industries and their workers because communication products are ephemeral, reducible to digital forms that make them among the most easily adaptable to new production processes. The current global transformation of the telecommunications industry offers a stark example. The development of digital systems provide a more cost-efficient means of producing, controlling, and managing signals, thereby permitting companies worldwide to eliminate 20% of the entire telecommunications workforce in a 10-year period. However one fashions an answer, political economy reminds that this is a political, not just a technical, process. As Harvey (1989) noted, "The same shirt designs can be produced by large-scale factories in India, co-operative production in the 'Third Italy,' sweat-shops in New York and London, or family labour systems in Hong Kong"(p. 187).

Globalization and industrial restructuring mutually influence patterns of government restructuring. Following on the work of Murdock (1990), one can identify four dimensions of changed-state activity. Commercialization establishes state functions such as providing postal and telecommunications services, principally along business or revenue-generating lines. Privatization takes this a step further by turning these units into private businesses and liberalization invokes the state's approval to opening markets to competition. Finally, internationalization links the state to other states to shift economic and political authority to regional systems, such as is spelled out in the North American Free Trade Agreement (NAFTA), and to international bodies, as laid out in the creation of a World Trade Organization through the General Agreement on Tariffs and Trade (GATT).

The political economy of communication addresses this process mainly in terms of the institutional extension of corporate power in the communication industry. This is manifested in the sheer growth in the size of media firms, measured by assets, revenues, profit, employees, or share value. Political economy has specifically examined growth by analyzing different forms of corporate concentration (Herman & Chomsky, 1988).

Horizontal concentration takes place when a firm in one line of media buys a major interest in another media operation that is not directly related to the original business, or when it takes a major stake in a company entirely outside the media. The typical example of the first, or cross-media concentration, is the purchase by a firm in an older line of media, say a newspaper, of a firm in a newer line of media, such as a radio or television station.

Vertical integration describes the amalgamation of firms within a line of business that extend a company's control over the processes of production and distribution. MCA's purchase of Cineplex-Odeon gave the former, a major Hollywood producer, control over a major film distribution company. This is also referred to as forward integration because it expands a firm further along the process. Backward vertical integration took place when *The New York Times* purchased paper mills in Quebec, thereby expanding the company down the production process.

Political economists of communication have focused significant attention on the extension of integration across borders as companies such as Time Warner, Bertelsmann, News Corporation, Matsushita, Hachette, Havas, Fininvest, and Sony developed into transnational conglomerates that rival, in size and power, firms in any industry. They are just beginning to take up the development of flexible forms of corporate power evidenced in the joint ventures, strategic alliances, and other short-term, project-specific "teaming arrangements" that bring together companies or parts of companies, including competitors. These take advantage of more flexible means of communication to unite and separate for mutual interest. They have brought together media firms such as Time Warner and the French Canal Plus, computer giants IBM and Apple, and telecommunication leaders AT&T, France Telecom, and Deutsche Bundesposte.

A start has also been made on political economic work that addresses the international division of labor and labor internationalism (Sussman & Lent, 1991). One consequence of spatialization is the development of global labor markets. Business can now take advantage of differential wages, skills, and other important characteristics on an international scale. Much of the early political economic work in this area concentrated on the spread of the hardware (Southeast Asia) and data entry (the Caribbean) businesses into the Third World, where companies were attracted by low wages, tax and other incentives, and authoritarian rule (Heyzer, 1986; Sussman, 1984). More recently, the scope of research has expanded to address capital's growing interest in looking to the less developed world for sources of relatively low wage but skilled labor, needed in such areas as software development (Yourdon, 1993), and also to the developed world, where a prime example is the growth of U.S. film and video production in Toronto, Vancouver, and other parts of Canada.

The growth of the international division of labor in communication has sparked an interest in labor internationalism in communication. This includes making use of the means of communication, including new technologies, to forge close links among working class and trade union interests across borders (Waterman, 1990). Again, like much of the literature on the commodification of labor in communication, it has only begun to address what is a central focus of attention in other fields of political economy. As it does so, the political economy of communication needs to resist the understandable tendency to view spatialization as just another term for globalization. Spatialization also leads to a political economic analysis of nationalism and other forms of localization. The world map is not only being redrawn to conform to changes in the space of global flows. Alongside globalization we find a resurgent nationalism (e.g., Russia) and of nationalisms within nationalism (e.g., Bosnia) that contribute to, and conflict with, tendencies at the global level. Finally, these local and nationalist processes also need to be situated against socialism, historically the principal alternative to a global capitalist political economy (Ahmad, 1992).

The third entry point is structuration, a process given recent prominence in Giddens' (1984) work. Structuration describes how human agency constitutes structures that provide the very "medium" of that constitution. This amounts to a contemporary rendering of Marx's notion, that people make history, but not under conditions of their own choosing. The term responds to concerns with functionalist, institutional, and structuralist approaches, arising out of their tendency to present structures as fully formed, determining entities. Specifically, it helps to balance a tendency in political economic analysis to concentrate on structures, typically business and governmental institutions, by incorporating the ideas of agency, social process, and social practice.

Concretely, this means broadening the conception of class from its structural or categorical sense, which defines it in terms of what some have and others do not to incorporate both a relational and a constitutional sense of the term. A relational view of class foregrounds the connections, for example, between capital and labor, and the ways in which labor constitutes itself within the relationship, and as an independent force in its own right. This takes nothing away from the value of seeing class, in part, as a zero-sum game that pits haves against have-nots.

The political economy of communication has addressed class in categorical terms by producing research that documents persistent inequities in communication systems, particularly access to the means of communication and to the reproduction of these inequities in social institutions (Golding & Murdock, 1991; Schiller, 1989). This has been applied to labor, particularly in research on how communication and information technology has been used to automate and deskill, rather than to enrich and, in Zuboff's (1988) term, "informate" work (Webster & Robins, 1986). It has also been used to show how the means of communication are used to measure and monitor work activity in systems of surveillance that extend managerial control over the entire labor process in precise detail (Clement, 1992).

Rethinking the political economy approach means expanding on this conception with, first, a relational view of class that defines it according to those practices and processes that link class categories. In this view, the working class is not defined simply by lack of access to the means of communication, but by its relationships of harmony, dependency, and conflict with the capitalist class. Moreover, a constitutional conception of class views the working class as producer of its own, however tenuous, volatile and conflicted identity, in relation to capital and independently of it. Political economists have responded tentatively to this call for a class constitutional approach, though a literature on the working class and labor has begun to address this area more forthrightly (see, e.g., Bekken, 1990; McChesney, 1993). The point is that there is a pressing need to examine oppositional and alternative class-based movements, ranging from revolutionary struggles in Latin America, the Caribbean, Asia, Africa, and Eastern Europe, where the mass media often stirred revolutions: to alternative media in the West, which provided a trade unionist, rank-and-file, or socialist alternative to capitalist "common sense." The point is not to engage in romantic celebration, but, at the very least, to demonstrate how classes constitute themselves, how they make history in spite of those well-researched conditions that constrain this history-making activity.

Rethinking political economy also means balancing another tendency in political economy: When it has given attention to agency, process, and social practice, it tends to focus on social class. There are good reasons for this emphasis. Class structuration is a central entry point for comprehending social life, and numerous studies have documented the persistence of class divisions in the political economy of communication. Nevertheless, there are other dimensions to structuration that complement and conflict with class structuration, including gender, race, and those broadly defined social movements, which, along with class, constitute much of the social relations of communication. Unlike other approaches, political economy has not been entirely silent on the issue of gender, although it typically addresses the subject as a dimension of social class relations. For example, it has done this in research on information technology and the international division of labor, which addresses the double oppression that women workers face in industries such as microelectronics, in which they experience the lowest wages and the most brutalizing working conditions (Wright et al., 1987).

Additionally, although communication studies has addressed the question of imperialism extensively, principally by examining the role of the media and information technology in its constitution, it has done so primarily to advance a sense of the world as class-divided or, although less frequently, as gender-divided, rather than to understand it as race-divided. Yet one does not have to focus on South Africa to recognize that racial divisions are a principal constituent of the manifold hierarchies of the contemporary global political economy, and that race, as both category and social relationship, helps shape access to national and global resources, including communication, media, and information technology (Ahmad, 1992; Sivanandan, 1990).

From this use of structuration theory, one might think about society as the ensemble of structuring actions initiated by social actors that shape and are shaped by class, gender, race, and social movement relations. According to this view, society exists, if not as a seamless, sutured whole, at least as a field on which various processes mutually constitute identifiable social relationships. It thereby rejects the poststructuralist view that the social field is a continuum of subjectivities produced by purely nominal processes of categorization. Accordingly, class, race, gender, and social movements are real, both as social relationships and as foci of analysis.

One of the major activities in structuration is the process of constructing hegemony, defined as what comes to be incorporated and contested as the taken-for-granted, common sense, natural way of thinking about the world, including everything from cosmology to ethics to everyday social practices. Hegemony is a lived network of mutually constituting meanings and values, which, as they are experienced as practices, appear as mutually confirming. For example, although political economy addresses agents as social rather than individual actors, it recognizes the significance of the hegemonic process of individuation. The concept, taken chiefly from Poulantzas (1978), refers to the practice of redefining social actors, capital, and labor, particularly, as individual subjects whose value is connected to individual rights, expression, the exercise of political responsibility in voting, and freedom of consumption. These actions, taken in the name of government but bound up with the exercise of class rule, isolate individuals from one another, from their social identities, and from those with the power to carry out individuation. One of the central tensions, conflicts, and struggles within the process of structuration is between social and individuating tendencies. In conclusion, out of the tensions and clashes among various structuration processes, the media come to be organized in their full mainstream, oppositional, and alternative forms (Williams, 1975).

Rethinking and renewing political economy also requires one to look outward at the relationship between the discipline and those on its borders. Although, admittedly, one can map the universe of academic disciplines in numerous ways, it is useful to situate the political economy of communication opposite cultural studies on the one side and policy studies on the other.

Cultural studies is a broad-based intellectual movement that concentrates on the constitution of meaning in texts, defined broadly to include all forms of social communication (During, 1993). The approach contains numerous currents and fissures that provide for considerable ferment from within. Nevertheless, it can contribute to the process of renewing political economy in several ways. Cultural studies has been open to a broad-based critique of positivism and an effort to build a more open philosophical approach that foregrounds the subjective and social constitution of knowledge. It has also aimed to broaden the sense of what comprises the substance of cultural analysis by starting from the premise that culture is ordinary, produced by all social actors, rather than mainly by a privileged elite, and that the social is organized around gender and nationality divisions and identities as much as by social class.

Although political economy can learn from these departures, it can equally contribute to rethinking cultural studies. Even as it takes on a philosophical approach that is open to subjectivity and is more broadly inclusive, political economy insists on a realist epistemology that maintains the value of historical research, of thinking in terms of concrete social totalities, moral commitment, and of overcoming the distinction between social research and social practice. It therefore departs from the tendency in cultural studies toward what Pêcheux calls "the narcissism of the subject," by correcting the now fashionable inclination to reject thinking in terms of historical practices and social wholes. Political economy also departs from the growing proclivity toward an obscurantism in cultural studies that belies the original view that cultural analysis should be accessible to those ordinary people who are responsible for its social constitution. Finally, it eschews the propensity in cultural studies to reject studies of labor and the labor process in favor of examining the social "production" of consumption and the ensuing tendency to reject labor as holding any value in contemporary movements for social change (Luke, 1989).

Political economy can also learn from the development of a policy science perspective whose political wing has tended to place the state at the center of analysis and whose economic wing aims to extend the application of primarily neoclassical economic theory over a wide range of political, social, and cultural life (Posner, 1992; Stigler, 1988). Political economy has tended to "read" the state and other "superstructural" forces from the specific configuration of capital dominant at the time and therefore benefits from an approach that takes seriously the constitutive role of the state. Moreover, political economy shares with policy science the interest in extending analysis over the entire social totality, with an eye to social transformation.

Nevertheless, political economy departs fundamentally from the policy science tendency toward a pluralist political analysis, which views the state as the independent arbiter of a wide balance of social forces, none of which holds sway. Against this, political economy insists on the power of capital and the process of commodification as the starting point of social analysis. Furthermore, political economy rejects the policy science tendency to build its analysis of the social totality and of those values that should guide its transformation on the basis of individualism and market rationality. Against this, it insists on social processes, starting from social class and labor, and on setting community and public life against the market and a rationality that actually reproduces class power.

REFERENCES

Ahmad, A. (1992). *In Theory: Classes, nations, literatures*. London: Verso.

Bekken, J. (1990, August). "This paper is owned by many thousands of working men and women": Contradictions of a socialist daily. Paper presented at the Annual Meeting of the Association for Education in Journalism and Mass Communication, Boston.

Benton, T. (1989). Marxism and natural limits: An ecological critique and reconstruction. *New Left Review*, No. 178, 51-86.

Berberoglu, B. (Ed.). (1993). *The labor process and control of labor: The changing nature of work relations in the late twentieth century*. Westport, CT: Praeger.

Boafo, S.T.K. (1991). Communication technology and dependent development in sub-Saharan Africa. In G. Sussman & J. A. Lent (Eds.), *Transnational communications: Wiring the Third World* (pp. 103-124). Newbury Park, CA: Sage.

Braverman, H. (1974). *Labor and monopoly capital*. New York: Monthly Review.

Brennan, G., & Buchanan, J. M. (1985). *The reason of rules: Constitutional political economy*. Cambridge, UK: Cambridge University Press.

Burawoy, M. (1979). *Manufacturing consent*. Chicago: University of Chicago Press.

Cardoso, F. H., & Faletto, E. (1979). *Dependency and development in Latin America*. Berkeley: University of California Press.

Carlyle, T. (1984). *A Carlyle reader* (G.B. Tennyson, Ed.). New York: Cambridge University Press.

Castells, M. (1989). *The informational city: Information technology, economic restructuring, and the urban-regional process*. Oxford: Basil Blackwell.

Clement, A. (1992). Electronic workplace surveillance: Sweatshops and fishbowls. *Canadian Journal of Information Science, 17*(4), 18-45.

Coase, R.H., & Barrett, E. W. (1968). *Educational TV: Who should pay?* Washington, DC: American Enterprise Institute for Public Policy.

Coase, R.H. (1991). *The nature of the firm: Origins, evolution, and development* (O. E. Williamson & S. G. Winter, Eds.). New York: Oxford University Press.

Douglas, S. (1986). *Labor's new voice: Unions and the mass media*. Norwood, NJ: Ablex.

During, S. (1993). *The cultural studies reader*. London: Routledge.

Edwards, R. (1979). *Contested terrain: The transformation of the workplace in the twentieth century*. New York: Basic.

Fishman, M. (1980). *Manufacturing the news*. Austin: University of Texas Press.

Foster, J. B. (1988). The fetish of fordism." *Monthly Review, 39*, 14-20.

Galbraith, J. K. (1985). *The new industrial state* (4th ed.). Boston: Houghton Mifflin.

Gandy, O. H., Jr. (1992, Summer). The political economy approach: A critical challenge. *Journal of Media Economics*, pp. 23-42.

Garnham, N. (1990). *Capitalism and communication: Global culture and the economics of information*. London: Sage.

Giddens, A. (1984). *The constitution of society: Outline of a theory of structuration*. Berkeley: University of California Press.

Giddens, A. (1990). *The consequences of modernity*. Stanford, CA: Stanford University Press.

Golding, P., & Murdock, G. (1991). Culture, communication, and political economy. In J. Curran & M. Gurevitch (Eds.), *Mass media and society* (pp. 15-32). London: Edward Arnold.

Hardt, H. (1990). Newsworkers, technology, and journalism history. *Critical Studies in Mass Communication, 1*(4), 346-365.

Harvey, D. (1989). *The condition of postmodernity*. Oxford: Basil Blackwell.

Herman, E. S. & Chomsky, N. (1988). *Manufacturing consent: Political economy of the mass media*. New York: Pantheon.

Heyzer, N. (1986). *Working women in Southeast Asia: Development, subordination, and emancipation*. Philadelphia: Open University Press.

Jevons, W. S. (1965). *The theory of political economy*. New York: A.M. Kelly.

Jussawalla, M. (1986). *The passing of remoteness: The information revolution in the Asia-Pacific*. Singapore: Institute of Southeast Asian Studies.

Keynes, J. M. (1964). *The general theory of employment, interest, and money*. New York: Harcourt, Brace & World.

Kraft, P., & Dubnoff, S. (1986). Job content, fragmentation and control in computer software work. *Industrial Relations, 25*, 184-196.

Lash, S., & Urry, J. (1987). *The end of organized capitalism*. Madison: University of Wisconsin Press.

Lebowitz, M. (1986). Too many blindspots on the media. *Studies in Political Economy*, No. 21, 165-173.

Lipietz, A. (1988). Reflections on a tale: The marxist foundations of the concepts of regulation and accumulation. *Studies in Political Economy, 26*, 7-36.

Luke, T. (1989). *Screens of power: Ideology, domination, and resistance in informational society*. Urbana and Chicago: University of Illinois Press.

Marshall, A. (1961). *Principles of economics*. London: MacMillan. (Original work published 1890)

Marx, K. (1973). *The Grundrisse: Foundations of the critique of political economy* (M. Nicolaus, Trans.). Harmondsworth: Penguin.

Marx, K. (1976). *Capital: A critique of political economy* (Vol. 1, B. Fowkes, Trans.). London: Penguin.

Massey, D. (1992). Politics and space/time. *New Left Review*, No.196, 65-84.

Mattelart, A., & Mattelart, M. (1992). *Rethinking media theory: Signposts and new directions* (J. A. Cohen & M. Urquidi, Trans.). Minneapolis: University of Minnesota Press.

Mattelart, A., & Siegelaub, S. (1983). *Communication and class struggle: Vol. 2: Liberation, socialism.* New York: International General.

McChesney, R. W. (1993). *Telecommunications, mass media and democracy: The battle for the control of U.S. broadcasting.* New York: Oxford University Press.

Miège, B. (1989). *The capitalization of cultural production.* New York: International General.

Morris-Suzuki, T. (1986). The challenge of computers. *New Left Review*, No. 160, 81-91.

Mosco, V., & Wasko, J. (Eds.). (1983). *The critical communications review. Vol. 1: Labor, the working class, and the media.* Norwood, NJ: Ablex.

Mosco, V., & Zureik, E. (1987). *Computers in the workplace: Technological change in the telephone industry.* Ottawa: Government of Canada, Department of Labour.

Murdock, G. (1978). Blindspots about Western Marxism: A reply to Dallas Smythe. *Canadian Journal of Political and Social Theory*, 2(2), 109-119.

Murdock, G. (1990). Redrawing the map of the communication industries. In M. Ferguson (Ed.), *Public communication: The new imperatives* (pp. 1-15). Beverly Hills, CA: Sage.

Murdock, G., & Golding, P. (1979). Capitalism, communication, and class relations. In J. Curran, M. Gurevitch, & J. Woolacott (Eds.), *Mass communication and society* (pp. 12-43). Beverly Hills, CA: Sage.

Nielsen, M. (1990). Labor's stake in the electronic cinema revolution. *Jump Cut*, No. 35, 78-84.

Owen, B. M., & Wildman, S. S. (1992). *Video economics.* Cambridge, MA: Harvard University Press.

Owen, R. (1851). *Labor: Its history and prospects.* New York: No publisher listed.

Posner, R. A. (1992). *Sex and reason.* Cambridge, MA: Harvard University Press.

Poulantzas, N. (1978). *State, power, and socialism.* London: New Left Books.

Resnick, S. A., & Wolff, R. D. (1987). *Knowledge and class: A Marxian critique of political economy.* Chicago: University of Chicago Press.

Roach, C. (Ed.). (1993) *Communication and culture in war and peace.* Newbury Park, CA: Sage.

Roncagliolo, R. (1986). Transnational communication and culture. In R. Atwood & E. G. McAnany (Eds.), *Communication and Latin American society* (pp. 79-88). Madison: University of Wisconsin Press.

Russial, J. T. (1989). *Pagination and the newsroom: Great expectations.* Unpublished doctoral dissertation, Temple University, Philadelphia, PA.

Schiller, H.I. (1989). *Culture, Inc.* New York: Oxford University Press.

Schiller, H.I. (1992). *Mass communication and American empire* (2nd ed.). New York: Augustus Kelly. (Original work published 1969)

Schumpeter, J. (1942). *Capitalism, socialism, and democracy.* New York: Harper and Brothers.

Sivanandan, A. (1990). *Communities of resistance: Writings on black struggles for socialism.* London: Verso.

Smith, A. (1937). *An inquiry into the nature and causes of the wealth of nations.* New York: Modern Library. (Original work published 1776)

Smith, A. (1976). *The theory of moral sentiments.* Indianapolis, IN: Liberty Classics. (Original work published 1759).

Smythe, D.W. (1977). Communications: Blindspot of Western Marxism. *Canadian Journal of Political and Social Theory, I*(3), 1-27.

Smythe, D.W. (1981). *Dependency road: Communication, capitalism, consciousness and Canada.* Norwood, NJ: Ablex.

Stigler, G. J. (Ed.). (1988). *Chicago studies in political economy.* Chicago: University of Chicago Press.

Sussman, G. (1984). Global telecommunications in the third world: Theoretical considerations. *Media, Culture, and Society, 6*, 289-300.

Sussman, G., & Lent, J. A. (Eds.). (1991). *Transnational communications: Wiring the third world.* Newbury Park, CA: Sage.

Tang, W. H., & Chan, J. M. (1990). The political economy of international news coverage: A study of dependent communication development. *Asian Journal of Communication, 1*(1), 53-80.

Veblen, T. (1932). *The theory of business enterprise.* New York: Scribners.

Veblen, T. (1934) *The theory of the leisure class.* New York: Modern Library. (Original work published in 1899).

Wallerstein, I. (1979). *The capitalist world economy.* New York: Cambridge University Press.

Waring, M. (1988). *If women counted: A new feminist economics.* New York: Harper Collins.

Wasko, J. (1983). Trade unions and broadcasting. In V. Mosco & J. Wasko (Eds.), *The critical communications review. Vol. 1: Labor, the working class, and the media* (pp. 85-113). Norwood, NJ: Ablex.

Wasko, J., & Mosco, V. (Eds.) (1992). *Democratic communication in an information age.* Toronto: Garamond and Norwood, NJ: Ablex.

Waterman, P. (1990). Communicating labor internationalism: A review of relevant literature and resources. *The European Journal of Communication, 15*(1/2), 85-103.

Waterman, P. (1992). The transmission and reception of international labour information in Peru. In J. Wasko & V. Mosco (Eds.), *The critical communications review. Vol. 1: Labor, the working class, and the media* (pp. 224-241). Norwood, NJ: Ablex.

Webster, F., & Robins, K. (1986). *Information technology: A Luddite analysis.* Norwood, NJ: Ablex.

Williams, R. (1975). *Television, technology and cultural form.* London: Fontana.

Wright, B. D., et al. (Eds.). (1987). *Women, work, and technology.* Ann Arbor: University of Michigan Press.

Yourdon, E. (1993). *The decline and fall of the American programmer.* Englewood Cliffs, NJ: Prentice Hall.

Zimbalist, A. (1979). Technology and the labor process in the printing industry. In A. Zimbalist (Ed.), *Case studies in the labor process* (pp. 103-126). New York: Monthly Review.

Zuboff, S. (1988). *In the age of the smart machine.* New York: Basic.

3

Information Networks: The Global Offshore Labor Force

Mark I. Wilson

The appeal of global markets and advances in transportation always led to a redrawing of trading patterns and routes, as relative costs of movement changed and new areas opened to wider markets. The impact of steam over sail, of rail over canal, and of refrigeration over deterioration shapes the production patterns of the industrial world. Transportation in a goods based world is central to production and

*This research has been funded by grants from the MSU Foundation and International Studies and Programs at Michigan State University, with additional support from the Institute for Public Policy and Social Research (MSU) and James Madison College Faculty Development Fund. Research has benefitted from the research assistance of Vincent Frillici. Oumatie Marajh conducted invaluable field research and data collection in the Caribbean.

The cooperation of government agencies in Ireland, Barbados, and Jamaica is greatly appreciated, as is assistance from many data-processing and telecommunications firms.

distribution, raising questions about the spatial organization of production in a services-based world. Telecommunications replaces tankers when the commodity is information, and the organization of production sensitive to distance, ocean, and mountain barriers may well be destabilized. Using telecommunications often minimizes distance and the physical barriers that impede movement, producing a global cost space quite different to land and sea transportation.

Facilitating and accompanying the development of telecommunications services is the rise of the information economy, although the growth of information-generating activities is only part of the changing structure of production. Information production is frequently directed to enhanced manufacturing and agricultural output, embodied in the technologies used for production. The development of the information economy is seen by the OECD (1993) to be driven by four processes: (a) growth in final demand for information products; (b) growing knowledge intensity of production in all sectors; (c) growth of public services such as education, health, and welfare; and (d) division of information labor into specialized information work. These forces shape not only domestic economies, but also the rapidly expanding globalization of production. With globalization comes new patterns of production, and in an information economy, the spatial context is information space, as found in Hepworth's (1990) description of the geographical setting for information generation, transmission, and use.

Passively, telecommunications and information technology can be seen as conduits for information flows, but they are also vehicles for control and access to reach databases, clients, and workers distributed across space. These elements are receiving overdue recognition by scholars with an interest in the political economy of information and the impact of technological change on production, spatial patterns, and individuals. Hepworth (1990) directs attention to the impact of computer networks and the role of information capital in manufacturing and the concept of information space. The uneven development stemming from applications of information capital is emphasized by Mulgan (1991) in his volume on the ways computer networks transform space. Kellerman (1993) adopts a spatial perspective on how telecommunications affects urban, regional, and international development. Castells (1989) emphasizes the social dimension of information technology in his view of the "dual city," with its social stratification and income polarization resulting from the restructuring of labor. The disparity Castells observes at the urban and metropolitan level is also a global phenomenon, as the information economy incorporates many nations and places.

Global services production captures the changing character and spatial context of capitalism far more dramatically than manufacturing because services' production and use of information allows the electronic movement of inputs far faster than is possible with manufacturing. Firms can specialize and adopt flexible production methods to conquer time, but time is still needed to move shoes from Asia to Europe, or appliances from Asia to North America. Harvey's (1989) concept of time-space compression reflected on centuries of production of physical output, yet this idea is no better illustrated than by the global service economy.

Information can be moved almost instantaneously around the world from producer to producer or end user. Space is conquered easily, or annihilated to use Harvey's expression, by telecommunications, in which distance is measured in seconds, satellite echo, or cents per minute. There is no example more telling of the compression of time and space as the case of an Ohio software firm. Its software code starts its day in the U.S. Midwest, and at the end of the day it is electronically transmitted to a branch facility in Hawaii, where an additional six hours of processing labor is performed. As the sun sets in Hawaii, the job is forwarded to yet another branch in Bangalore for another day's work before being returned at dawn to Ohio. Around the world in 24 hours, with 17 hours of work performed, there is no more dramatic example of the lengths to which production can and will be organized.

Globetrotting software code is but one of many tasks now performed almost without regard to distance by telecommunications-dependent production systems. It is important to make the distinction between distance and place, for although distance may well be compressed or annihilated by electronic media, the qualities of place remain more important than ever. The reason conquering distance is so important is to gain access to places with low-cost production opportunities. Places carry social, cultural, economic, and political characteristics that reward the effort needed to establish electronic and commercial linkages. The international division of labor developed because of place-specific variation in wages, working conditions, and labor markets. If places were not very different, then the need to conquer distance would be minimized. Of course, electronic access may well be the means to diminish difference, to be the force that thrives on place specific characteristics at the same time that it becomes a force to minimize them.

This chapter builds on themes of time-space compression, services globalization, and international division of labor by addressing modern firm behavior. Firms combine telecommunications with information technology to restructure their organization of service work, functionally and along spatial dimensions. The spatial impact of telecommunications is already evident for business services such as banking, software development, and information processing, but what are the larger questions and issues that these industries' experience suggest? Two aspects of the impact of communications on the spatial organization of services production are discussed: First, the way telecommunications use space creates a significantly different production geography than existed previously; and second, how telecommunications advances led to global services production and the international division of information labor.

The impact of declining communications costs and expanded capacities is well illustrated by a case study of the globalization of information-processing services, the phenomenon of offshore back offices. Back offices collect, manage, and process information as an intermediate input to the production of goods and services. Activities commonly undertaken by back offices include data entry and processing, database management, accounting and financial services, processing of magazine subscriptions and insurance claims, and computer software development. Over the past decade, these services have started to leave major cities in developed economies to locate in less costly locations in urban and rural areas of developing economies. Global telecommunications systems allow these back offices to function as part of worldwide networks that, in terms of firm operations, minimize the impact of space.

The forces operating to reshape the production cost map are: (a) firms seeking low cost labor and work environments, (b) public- and private-sector enhancement of infrastructure quality and capacity, and (c) public policy encouraged by, and directed toward, the location decisions of transnational firms. These forces are exemplified using case studies of information processing in Asia, Europe, and the Caribbean. Empirical data were collected by surveys and interviews of information processing firms, communications organizations, and government agencies and policymakers.

TELECOM TECTONICS

In a service economy operating on information generation and flows, the global cost aspect is decided not as much by shipping or aviation as by telecommunications. The economic advantages of global electronic access to reach markets and low-cost labor drives the continuous development of telecommunications capacity. As telecommunications costs decline, the global cost space for telecommunications dependent producers changes also. New opportunities arise for firms to restructure production and its location while lowering operating costs. Information flows and telecommunications access is not unfettered, however, as access to advanced technology depends on a willingness to invest and the availability of capital, as well as on politically determined rights for telecommunications access.

Metaphorically, continents and countries are moving closer together in telecommunications cost space, with Britain moving from being over $1 per minute away from the United States to being only 20 cents per minute away. Also, just as continental drift brought great physical change, telecom tectonics is bringing social and economic change. In telecom tectonics terms, not all countries and locations are moving at the same pace. Many developing countries unable to afford electronic technology remain in place or lose ground, or they have major cities centrally placed in information space and hinterlands far distant because of outmoded domestic telecommunications systems. Similarly, parts of Manhattan may well be integral to the information world, whereas blocks away such connectivity may be irrelevant or impossible.

Growth and cost reductions in telecommunications capacity have facilitated increased flows of information and more rapid dissemination of knowledge. The real cost of international telecommunications services has declined remarkably over the past decade. Between 1982 and 1992, the real cost of a three minute standard call between many major OECD countries has declined by up to one half. For example, a three-minute telephone call from the United States to France decreased 54%, from $7.31 (1992 $) in 1982 to $3.83 in 1992, with similar levels of cost reduction for calls from the United States to many countries in Europe. Telephone calls to Asia decreased, in real terms, by 50% to 60% over the same decade. Not only has the cost of telecommunications declined, but the amount of information that can be transferred has increased. Although voice transmission has not been enhanced greatly, fax machines and international computer file transfer have increased information flows dramatically. The changing cost structure of international telephone calls from the United States is shown in Table 3.1.

The reduction of real telephone costs has not been the same for all countries, with PTTs (state-run postal, telegraph, and telephone systems) in many countries slow to pass on the cost savings brought by advances in telecommunications technology. The privatization of PTTs in some countries, however, has hastened competition for international calls and brought reduced telephone charges. For example, between 1982 and 1992, telephone calls from the United States to Hong Kong, which is privatized, declined 41%, from $10.73 to $6.32, whereas a similar call to China, which is state-run, declined only 30%, from $12.42 to $8.74.

Table 3.1. International Telephone Charges for 3-Minute Standard Call to Selected Countries from the United States.

	1982 $	1992 $	Percent Change
Australia	10.73	5.60	-48%
France	7.31	3.83	-46%
Germany	7.31	3.95	-46%
Italy	7.31	3.96	-46%
Hong Kong	10.73	6.32	-41%
Japan	10.73	5.53	-48%
Singapore	10.73	6.02	-44%
South Korea	10.73	6.42	-40%
Canada[a]	3.10	1.75	-44%
Jamaica	5.73	3.46	-40%
Mexico[a]	3.36	1.26	-63%

[a]Rate for call up to 1000 miles
Source: Federal Communications Commission, *Statistics of Communications Common Carriers* [Annual] 1982-1992.

Also, the telecommunications cost surface depends on the source of the call, with reductions from one country to another not automatically matched for the reverse call. For example, a standard three-minute telephone call from the United States to Japan costs $5.53, whereas a call in the other direction cost 16% more ($6.30) in 1992. Some of these differences can be accounted for by currency levels, but they are also caused by limited competitiveness and monopoly power by PTTs.

Advanced telecommunications facilities are no longer the realm of leading economies. During the 1980s, Ireland completely renovated its telephone system with a digital network. Barbados' digitalized communications system offers direct international dialing for telephone/fax and satellite-based, high speed data transmission capacity to North America and Europe. Currently, Pakistan is implementing a national fiber-optic core telecommunications system to link major centers. Declining costs of information technology open new opportunities for trade and employment, but also carry opportunity and social costs. Market pressures on firms and government can lead to substantial infrastructure investment to gain a globally competitive position.

GLOBALIZATION OF SERVICES

The international spread of corporate services reflects a number of trends, all of which demand a sophisticated telecommunications infrastructure. First, firms grow as facilitators of information exchange such as financial data, news, and entertainment networks. Second, they accelerate the pace of economic activity to gain additional profit such as international financial and banking flows and stock market operations (see Hepworth, 1990; Langdale, 1991; Warf, 1989). Third, firms offer a broader range of services, through networks of networks like automated teller machines. Fourth, they expand markets to new areas such as banking, travel, or reservations services, and, finally, they enhance domestic production by using telecommunications to access overseas low cost workers and inputs such as the offshore back office (Wilson, 1995a).

Transnational corporations operate across many countries and cities to develop and exploit the production possibilities of telecommunications and information technology. Central to global operations for many firms is access to low-cost labor to serve international markets. Examples of this corporate form include airlines and hotel services, specialized retailing, and some financial services. Global operations necessitate, and are made possible by, global telecommunications systems to link all components of the corporation.

In the United States, back office workers earn $7 to $10 per hour, plus benefits in many cases. Data entry workers earn an average of $7.85 per hour, with senior workers averaging $9 per hour. Wages are highest in the West and lowest in the Great Lakes states (AMS, 1991). Firms moving back office tasks offshore can save as much as 50% over U.S. labor costs, with average cost savings of at least one third. Woodward (1990) analyzed international data entry costs by equating input costs to output of verified keystrokes. This approach takes into account differences in efficiency across locations, with some areas using low-cost workers in labor-intensive, double entry systems (Asia), whereas more advanced locations use better educated workers to maintain efficiency (Ireland). Comparisons show lowest costs in China, the Philippines, and the Caribbean, oftentimes 50% to 100% less than some European countries, the United States, Canada, and Japan. In both U.S. and offshore back offices, labor is primarily female and often drawn from the secondary labor market.[1]

---

[1]The relationship between women and technology in female-oriented clerical tasks shapes the production and, therefore, the location of this economic activity; for example, Andolsen (1989), Barker and Downing (1985), Carter (1987), Lowe (1987).

Once connected in a global network, firms have a range of service options. They can concentrate expertise in low cost or highly efficient locations while maintaining service throughout the system such as airline central reservations systems. They can gain economies of scale by operating at a global, rather than local, level so that telephone calls from clients can bypass a busy or closed local office and be handled elsewhere. For example, when closed, one international car rental corporation in Australia forwards telephone calls to its Tulsa (U.S.) reservations center. As another example, Richardson (1994) notes British Airways' switching of telephone traffic between Britain and the United States, using spare capacity on its data lines.

THE GLOBAL BACK OFFICE

Production costs are lowered and productivity increased by changing the ways that services are produced and by relocating or outsourcing to less costly areas. Service tasks are simplified and made routine so that less skilled and costly labor can be used or the process becomes fully or partially automated. The routinization of services production often requires previously bundled operations to be broken into simpler functions. Each function can then be processed in isolation, often by automated means or workers focusing on a narrow range of tasks. Instead of needing workers with a range of abilities to work on multiple purpose tasks, simplification establishes mass production of services, with each worker focusing on one or several specific activities. The positive and negative impacts of deskilling on workers and productivity are discussed in Braverman's (1974) analysis of production methods. Central to the deskilling process within the compelling momentum of capital and technological development is the tradeoff of greater productivity against worker alienation, isolation, and income stratification.

An additional dimension of deskilling is the ability to spatially unbundle production, separating tasks and placing them in the most productive location. The unbundling of tasks allows functions historically carried out in one place or division of a firm to be moved elsewhere. Back office activities can be unbundled and moved because they do not require much interaction with other units of the firm, and any interaction needed can be effectively achieved through telecommunications and computer networks. The cost advantages of distant locations counter any intrafirm communications disadvantages.

The development of global back offices is best portrayed as an extension of domestic trends in many countries, with clerical tasks being relocated to low-cost locations away from city and high value-added head office functions. The three spatial components of back offices are suburban, regional, and offshore operations. Each shows a similar pattern, with telecommunications central to the ability to access low cost locations while still remaining sufficiently connected to the firm or client.

A common and long-standing form of back office relocation is from corporate headquarters in the central business district (CBD) to surrounding suburbs of the city. Daniels (1987) and Marshall (1988) emphasize the use of telecommunications to link costly CBD locations to less costly suburban facilities. Suburban locations offer access to a primarily female labor market, characterized by lower wages and better educated part-time workers showing greater productivity for repetitive tasks (Baran, 1985; Metzger & Von Glinow, 1988; Moss & Dunau, 1987; Nelson, 1986). In her study of metropolitan Washington, DC, Meany (1991) emphasizes the wide range of business activities being decentralized. The spatial division of labor for information workers in cities and metropolitan areas reflects the spatial and social restructuring that also characterizes back offices at regional and international levels.

At the national level, firms use telecommunications to produce services at low-cost locations such as airline and hotel reservation services, payroll and financial services, and data processing. Regional back offices appeal to firms with national and international sales and multiple locations and with minimal need for physical interaction between headquarters and back office. Just as with suburban back offices, regional facilities emphasize the ability of firms to divorce back office service functions from their other activities to relocate in search of low-cost labor and operations.

Regional back offices are exemplified by a number of major corporate relocations. Warf (1989) profiles the move by American Express of its credit card processing facilities from New York to Florida and Arizona in order to cut costs, whereas Posthuma (1987) cites Citibank's shift of data-processing activities from New York to South Dakota. Metropolitan Life Insurance, which has an offshore facility in Ireland, also has data-processing centers in Greenville, SC; Scranton, PA; and Wichita, KS (Moss & Dunau, 1987). More recently, several major back offices have been established in Jersey City, NJ by firms in New York: First Chicago Trust moved 1,000 jobs from New York to Jersey City to lower costs of its dividend check-processing operations (Barsky, 1992), whereas Merrill Lynch transferred 2,500 back office jobs to Jersey City, and Lehman Brothers moved 900 back office workers from Manhattan (Deutsch, 1993).

Offshore back offices are an extension of suburban and regional back offices. Firms in advanced countries and high-cost locations send data- and information-processing tasks to low-cost sites abroad, predominantly in developing countries. Significantly, offshore offices frequently produce services that are inputs to domestic production and that serve domestic demand. Offshore locations are chosen not to serve foreign markets or to expand operations, but solely to provide cost effective inputs to production in the source nation, often corporate headquarters.

Offshore offices are organized either as offshore firms operating under contract or as subsidiaries of transnational firms. The problems of coordinating production with contract firms offshore led to the emergence of agencies that connect source firms to contract back offices offshore. Agencies are a response to the considerable information and organizational costs of establishing a contract offshore office. Lack of knowledge about offshore operators, the potential for delays and errors, and misunderstandings about terms and conditions slowed the growth of contract offices. Agencies contract directly with domestic firms for office tasks and then subcontract work to a range of offshore producers in a number of countries, depending on current demands and workloads. More recently, offshore offices are emerging as direct subsidiaries. Corporations establish offshore offices that are dedicated to serving internal requirements for office services. In most cases, the offshore office is indistinguishable from domestic operations in terms of organization and interaction with other units of the firm.

The operation of offshore back offices is similar for most firms in the industry. Materials, usually in the form of documents or magnetic tapes, are sent by air to processing facilities offshore. After processing, often done in less than a week and sometimes in two or three days, results are returned to originating data processing locations by courier, air freight, dedicated line, satellite, or public telephone line. The operational scale is substantial. Neodata handles over 200,000 transactions nightly between Ireland and Boulder, CO. Offshore back offices are well exemplified by case studies of airlines and Irish and Jamaican back offices.

Airline Back Offices

American Airlines is the pioneer in back office location, moving its data-processing operations from Tulsa, OK in the United States to Barbados in 1983. American Airlines assembles accounting material and ticket coupons in Dallas for transport on its own scheduled flights to Barbados for processing by its offshore subsidiary AMRIS (AMR Information Services/Caribbean Data Services). In Barbados, details of 800,000 American Airlines tickets are entered daily into a computer system and the data returned by satellite to its data center in Tulsa.

Data entry workers in Barbados earn US$2 to $3 per hour, with trainees earning less (US$1.50 per hour) and supervisors more (US$4 to $6). In addition, employers share with the employee the cost of the National Insurance and Social Security Scheme to cover pensions, sickness, invalidity, maternity, and unemployment benefits. Pressures on wages in Barbados are low because of high rates of unemployment (10% to 20%) in a highly literate workforce. Workers have the right to unionize, but there are no unions in back offices as the data processing industry actively discourages them. Posthuma (1987) reports that one firm used the threat of relocation to another low-wage Caribbean nation to prevent unionization.

Airline back offices are becoming increasingly sophisticated as more tasks are moved offshore. Carey (1992) reports that, in 1987, Swissair established a data input facility in Bombay to correct reservations errors and return the information to Zurich; Singapore Airlines planned a computer software and data-processing facility in the same building in Bombay. Cathay Pacific Airlines moved its accounting tasks to Guangzhou and its computer center to Sydney. For all these airlines, the central motivation is labor costs, primarily because of labor shortages and high costs in Hong Kong, Singapore, Zurich, and Tulsa and high land prices in Hong Kong. The Bombay operations of Singapore Airlines and Swissair offer substantial savings over wages and operating costs near headquarters.

Irish Back Offices

Over a dozen U.S. firms serve their internal needs for information processing through back offices located in small towns across western Ireland. The most common back office activity is insurance-related data/information processing, with facilities operated by Atlantic Run-Off (Galway), Beauman International (Dublin), Cigna Corporation (Loughrea), Massachusetts Mutual Life (Tipperary Town), Metropolitan Life (Fermoy), New York Life (Castleisland), and Travelers of Ireland (Limerick). Other back office operations include Neodata, a Dun and Bradstreet information processing subsidiary (four locations in and around Limerick); McGraw-Hill (Galway) for management of its magazine subscriptions database; and Stellar Systems (Galway) data processing for the international reinsurance industry. With the exception of Beauman International, located in Dublin, all facilities are in western Ireland.

The scale of back office operations in Ireland is considerable, with an initial investment frequently of one million Irish pounds (approximately US$1.6 million). Back offices undertake millions of transactions weekly. For example, McGraw-Hill handles over two million transactions annually for *Business Week,* one of 22 magazines it tracks at its Irish back office.

The primary attraction of Ireland is its well-educated, English-speaking workforce and wage rates at two thirds the prevailing U.S. rate. In addition, Irish workers cost less in terms of benefits and have lower turnover rates due to the scarcity of employment. The labor market advantages firms gain also relates to their location within Ireland. Many firms rely on overnight delivery of materials by air to Shannon Airport, outside Limerick, on the west coast of Ireland. Rather than locate near the airport, however, firms choose small towns with limited employment opportunity and high levels of unemployment; in Castleisland, New York Life received 600 applications for its first 25 positions. Firms often gain a monopoly advantage in such towns as major employers of high school leavers. Location choices guarantee access to educated workers without wage pressure from competing employers and provide firms with greater influence in local labor markets.

Jamaican Back Offices

Jamaica is one of the leading back office locations in the Caribbean. Pearson (1993) notes that some back office activity dates to the mid-1960s, but the logistics of this operation did not encourage the industry. It was not until advances in telecommunications and air transportation in the 1980s that viable back office operations developed. Back offices frequently work two to three shifts daily with workers averaging 7,000 keystrokes hourly at 99% accuracy. JAMPRO, the government economic development agency, estimates back office exports in 1990 at a daily capacity of 30 million keystrokes. Wages in Jamaica's back offices range between J$10 to $17 (US$0.50 to $0.85) per hour; with benefits added, costs average J$20 to $30 (US$1.00 to $1.50) per hour. The range of wages reflects the tasks undertaken, with the lowest wages paid to occasional and home workers and higher wages to skilled operatives.

The industry employs mainly clerical workers, although other occupations include managers, supervisors, and, occasionally, computer programmers and software developers. Clerical workers are expected to have a high school education and often some postsecondary technical training. The supply of experienced workers is assisted by the provision of data processing training in schools. Conversely, institutionalizing the supply of clerical workers may also depress wages in the future. With unemployment rates exceeding 25%, with even higher rates for women (Bourne, 1988), back office employment is seen as desirable by workers, even though the wages paid offer only low standards of living in

Jamaica's volatile economy. Antrobus (1989) is concerned that poor working practices and antiunion pressure exploits female labor in back offices and similar occupations.[2]

OFFICE POLITICS

Public policy and offshore back offices are closely related in several ways and in many countries. First, it is the international system of trade in producer services, using aviation and telecommunications systems, which allows offshore offices to exist. The ability of firms to establish international networks depends on the laws regulating trade and commerce. One reason that the United States is such a major source of offshore offices is that it does not limit the international movement of data and information. This contrasts with many countries in Western Europe that significantly limit transborder data flows or support public telecommunications systems that maintain high prices.

Second, public policy also influences production costs to attract offshore services that contribute to local employment and economic growth. Governments provide an extensive range of financial incentives and relocation assistance packages to attract firms and employment. The desirability to many economies of service employment has led to accelerating public support, even to the point at which attraction costs may equal or exceed benefits. Ireland, Jamaica, and Barbados offer the most advanced policies available to attract and develop offshore activities, although other countries provide some assistance but less comprehensive policies.

Ireland actively seeks service activities of foreign firms through tax benefits, financial incentives, labor/manpower planning, and the upgrading of communications facilities. Foreign firms manufacturing or producing services in Ireland pay corporate taxes at a rate of up to 10% compared to the standard rate of 43%; a benefit recently extended to 2010 (1980 Finance Act and 1990 Finance Bill). Tax benefits in general are used to attract firms to Ireland, with local taxes used to influence internal location to areas most in need, usually counties outside Dublin. Firms are also able to easily repatriate profits and face few restrictions on capital flows. European Union rules allow Ireland to offer more benefits than competing locations such as Scotland because of its need for development and to overcome its peripheral location in Europe. Policy is also designed to develop skills for workers, thus bridging the

---

[2]For additional information on the Caribbean back office, see Pearson (1993), Pearson and Mitter (1993), and Wilson (1995b).

gap between what is available in the labor market and the training that firms want. Even funding to send workers to the United States for training is available. Assistance is also provided for construction of facilities or rental premises, and capital grants are awarded for the purchase of computers, office equipment and furniture, and buildings.

Barbados offers low tax rates to offshore service firms, with no withholding tax on dividends, interest, and royalties. Firms are exempt from exchange controls and import duties and do not require licensing or local incorporation to operate. Jamaica also encourages offshore development, principally through its expansion of telecommunications capabilities and the establishment of a free trade zone and teleport facility in Montego Bay. Other countries in the Caribbean provide some fiscal incentives and also enjoy the benefits of the Caribbean Basin Initiative and its guaranteed access to the U.S. market. In Asia, offshore offices are predominantly a local phenomenon with less public policy direction. Leading industrialized nations such as Singapore, Taiwan, and South Korea focus public policy on advanced services rather than back office activities.

IMPLICATIONS

Offshore back offices illustrate how telecommunications can be used to access low cost labor overseas to produce intermediate inputs to the production of domestic goods and services. Although a relatively specialized phenomenon, offshore offices are instructive because they indicate the potential future for location of services. Declining real costs for telecommunications and increasing capacity for information transfer radically change the way information is handled and, given the significance of information to production, how the global economy is structured economically, socially, spatially, and politically. Part of this restructuring is the division of information labor, with services connected globally by telecommunications networks in order to establish facilities in the lowest cost locations.

Back offices exemplify two observations made about firm behavior and location by Storper and Walker (1989). First, the passive firm choosing from a set of preexisting conditions is, in fact, increasingly able to demand concessions from government or parallel location by suppliers. The economic power of many facilities allows the creation of advantageous economic space where none existed before; back offices offshore often are able to extract valuable concessions from governments eager for the many entry-level jobs they provide. Second, the social division of labor offers firms an economic advantage by relocating to gain workforce flexibility and a revised social contract. Despite limitations, back offices are showing some hallmarks of flexibility such as the shifting of risk away from firms onto workers and government and preferences for less regulated labor markets.

Back offices are a force to decentralize basic and clerical tasks to low-cost and less urban environments. Kellerman (1993) cites this view as the antipolitan argument, with telecommunications causing cities and metropolitan areas to lose their locational advantages for some activities. Employment has moved from central business districts to suburbs, smaller centers, and towns and to cities and towns in developing countries. Employment lost from U.S. cities such as New York and Tulsa has located in smaller centers such as Ft. Lauderdale, Phoenix, and urban areas in Barbados, Jamaica, and the Philippines. Internationally, employment has also moved to very small towns such as Castleisland, Fermoy, or Galway in western Ireland.

The attraction of basic service employment is appealing to many governments, especially those facing high rates of unemployment and slow rates of growth. Although public policy is commonly used to attract firms, it is important to note that the large number of potential sites for offshore offices means that many countries could be competing with one another to attract firms. In fact, the International Monetary Fund is actively encouraging Caribbean nations to develop their services exports through incentives to foreign firms. This produces the potential for strong competition for offshore offices by many countries, which erodes the net benefit to the successful location. The desirability of service employment is often gained only through the expenditure of taxpayer revenue on enhanced infrastructure and forgone tax revenue on back office operations. Such policies represent a shift of resources from local workers to foreign firms, which, in moderation, can greatly assist an economy, but when excessive can be costly and inequitable.

The future potential for offshore offices is not yet clear. On the one hand, technological advances can erase many of the tasks now carried out overseas. Some tasks initially undertaken offshore have returned to the United States for more efficient electronic processing. For example, Mead Data Central no longer processes documents in Asia for its online databases as it relies on tapes purchased directly from newspapers and printers. On the other hand, facilities such as those in Ireland and image processing in Jamaica represent advanced forms of services production offering more than just routine data entry. Information processing for the insurance industry requires educated workers with decision-making functions, which suggests that even more advanced tasks can be performed offshore in the future.

REFERENCES

Andolsen, B.H. (1989). *Good work at the video display terminal*. Knoxville: University of Tennessee Press.

Antrobus, P. (1989). Gender implications of the development crisis. In G. Beckford & N. Girvan (Eds.), *Development in suspense* (pp. 145-160). Kingston, Jamaica: Association of Caribbean Economists. .

Administrative Management Society (AMS). (1991). *Office, professional, and data processing salaries report*. Washington, DC: Author.

Baran, B. (1985). The technological transformation of white-collar work: A case study of the insurance industry. In H. I. Hartmann, R. E. Kraut, & L. A. Tilly (Eds.), *Computer chips and paper clips* (pp. 25-62). Washington, DC: National Academy Press.

Barker, J., & Downing, H. (1985). Word processing and the transformation of patriarchal relations of control in the office. In D. MacKenzie & J. Wajcman (Eds.), *The social shaping of technology* (pp. 147-164). Milton Keynes, UK: Open University Press. .

Barsky, N. (1992, September 28). First Chicago Corp. unit opts to move 1,000 jobs to New Jersey from New York. *Wall Street Journal*, B7.

Braverman, H. (1974). *Labor and monopoly capital: The degradation of work in the twentieth century*. New York: Monthly Review Press.

Bourne, C. (1988). *Caribbean development to the year 2000* London/Georgetown: Commonwealth Secretariat/Caribbean Community Secretariat.

Carey, S. (1992, November 30). Airlines seek to cut back-office costs by establishing offshore operations. *Wall Street Journal*, B5.

Carter, V. (1987). Office technology and relations of control in clerical work organization. In B. D. Wright (Ed.), *Women, work, and technology: Transformation* (pp. 202-291). Ann Arbor: University of Michigan Press.

Castells, M. (1989). *The informational city*. Oxford, UK: Basil Blackwell.

Daniels. P.W. (1987). Technology and metropolitan office location. *Service Industries Journal, 7*, 274-291.

Deutsch, C. H. (1993, October, 14). Lehman Bros. to Move 900 back-office jobs to Jersey City. *New York Times*, p. B6.

Harvey, D. (1989). *The condition of postmodernity*. Oxford, UK: Basil Blackwell.

Hepworth, M. (1990). *Geography of the information economy*. New York: Guilford.

Kellerman, A. (1993). *Telecommunications and geography*. London: Belhaven Press.

Langdale, J. (1991). Telecommunications and international transactions in information services. In S. Brunn & T. Leinbach (Eds.), *Collapsing time and space* (pp. 193-214). London: HarperCollins/Academic.

Lowe, G.S. (1987). *Women in the administrative revolution: The feminization of clerical work* . Toronto: University of Toronto Press.

Marshall, J.N. (1988). *Services and uneven development*. Oxford, UK: Oxford University Press.

Meany, J. (1991). Back office location in the Washington D.C. metropolitan area. In C. Fuchs, S. Mastran, & J. Meany (Eds.), *Office space, interchanges, and back-offices* (pp. 59-101). College Park MD: University of Maryland Geographical Publications.

Metzger, R.O., & Von Glinow, M. (1988). Offsite workers: At home and abroad *California Management Review* , *30*(3), 101-111.

Moss, M.L. & Dunau, A. (1987). Will the cities lose their back offices? *Real Estate Review* , *17*(1), 62-68.

Mulgan, G.J. (1991). *Communication and control*. New York: Guilford.

Nelson, K. (1986). Labor demand, labor supply, and the suburbanization of low wage office work. In A. J. Scott & M. Storper (Eds.), *Production, work, territory* (pp 149-171). Boston: Allen & Unwin.

Organisation for Economic Cooperation and Development (OECD). (1993). *Usage indicators: A new foundation for information technology policies*. Paris: Author.

Pearson, R. (1993). Gender and new technology in the Caribbean: New work for women. In J. Momsen (Ed.), *Women and change in the Caribbean* (pp. 287-295). Bloomington: Indiana University Press..

Pearson, R., & Mitter, S. (1993). Employment and working conditions of low-skilled information processing workers in less developed countries. *International Labour Review, 132*, 49-64.

Posthuma, A. (1987). *The internationalization of clerical work: A study of offshore office services in the Caribbean* (Science Policy Research Unit Paper 24). Brighton, UK: University of Sussex.

Richardson, R. (1994). Back officing front office functions: Organizational and locational implications of new telemediated services. In R. Mansell (Ed.), *Management of information and communications technologies: Emerging patterns of control* (pp. 309-335). London: Association for Information Management. .

Storper, M. & Walker, R. (1989). *The capitalist imperative*. New York: Basil Blackwell.

United States Federal Communications Commission. *Statistics of communications common carriers* (Annual). (1982-1992). Washington, DC: FCC.

Warf, B. (1989). Telecommunications and the globalization of financial services. *Professional Geographer, 41*(3), 257-271.

Wilson, M. (1995a). The office farther back: Business services, productivity, and the offshore back office. In P. Harker (Ed.), *The service and productivity challenge* (pp. 203-224). Boston: Kluwer.

Wilson, M. (1995b, Spring). Jamaica's back offices: Direct dial development or dependency? *Caribbean Geography*, pp. 117-126.

Woodward, P. (1990, January/February). Getting a start in data entry. *China Business Review*, pp. 20-23.

4

The Caribbean
Data Processors

Ewart C. Skinner

This chapter is a descriptive critical overview of the main features of the emergent Caribbean offshore data processing/data entry (ODP and ODE)[1] industry and its position in the new international division of labor (NIDL). It aims to draw a framework from which industry impact on the Caribbean worker can be assessed. It focuses on the political, economic, and sociohistorical aspects of the development of ODP and ODE in the Caribbean and their role in the NIDL. Four examples of aspects of Caribbean linkages to the data processing conduit are

[1]Data entry, the focus of this chapter, is a subsector of the data-processing industry. The U.S. Department of Labor, Bureau of Labor Statistics has revised the Standard Industrial Classification (SIC) index to reflect changes in the computer-related sector. Within its four-digit classification, the data-processing and preparation services occur under the (SIC 7374) category. The abbreviation ODE will be used for Offshore Data Entry and ODP for Offshore Data Processing when convenient.

provided, followed by a critical summary of the industry and recommendations of issues for further analysis.

The internationalization of data entry work stems from the convergence of telecommunications with computers, which allows corporate back-office, data entry chores to be outsourced, that is, exported for processing and subsequently electronically returned to the home office. According to Pearson and Mitter (1993), by the early 1970s, there was already in Jamaica and elsewhere in the Caribbean an "international relocation of tasks associated with mainframe computer processing, such as card punching" (p. 54). The Caribbean's continuing integration with the NIDL accelerated in the 1980s, "with offshore data entry . . . [and] the development of personal computer systems and electronic links to mainframe systems" (p. 55). An increasing percentage of data entry work is now sent to low-cost labor areas, increasingly offshore, which results in major savings for the base firm. This form of transfer follows the traditional economic trend of moving low-end manufacturing jobs to offshore sites in less developed countries (LDCs).

Optimistically termed the "sunrise sector," the industry represents, par excellence, the logistics of globalized information work—work that is moved through and among nations offering "competitive incentives, [adequate] utilities and support services" (The Services Group, n.d.).[2] Labor-intensive "high tech" combines the most basic human value (cheap labor) with the most advanced technology.

If the 1990s become known as the Caribbean "age of telecommunications," the watershed 1980s will have provided the critical impetus for its transformation; its catalyst, the telecommunications sector, is expected to provide lower end, "new opportunities for export diversification based upon cheap, efficient labor in information processing services" (The Service Group, n.d.). Optimistic assessments of industry growth by Caribbean development experts are based on the rate of adoption of data control electronics in international processing sites, where labor costs are competitive and latent data-processing talent is abundant. As technologically inevitable as it may be, important socio-economic and historical contexts underlie the industry's development.

The reestablishment of U.S. hegemony, Caribbean demand, transnational corporate (TNC) economic opportunity, and resurgent international faith in telecommunications as an economic multiplier all contribute to the industry's optimistic evaluation. With the promise of

---

[2]The quote is taken from The Services Group (TSG) brochure describing the firm and explaining the firm's services. TSG is a consulting firm that specialized in feasibility studies for nations interested in investing in international data services.

economic prosperity, data-processing enterprises set up shop in Jamaica, Barbados, the Dominican Republic, St. Lucia, and other island nations. Slow economic growth, high unemployment, and other symptoms of economic regression in the 1980s forced these nations to reconsider development strategies rejected in the region's industrial past. To put their economic houses in proper order, they moved with no-nonsense, economic pragmatism to reconstruct the offshore, export processing sector.[3] Anthony Maingot (1993) explained the pragmatic context that inspired offshore Caribbean enterprises:

> There can be no doubt about the compelling logic in this development thrust, a logic made more evident by the question: What were the alternatives, not theoretical or ideological, but actual and practical? A kind of economic law of necessity increasingly exercised a strong influence on development thinking, especially in the insular Caribbean. Decades of excessively ideological and partisan attacks on any offshore activity, whether tourism, medical schools, free trade zones, or even straight forward foreign investment, exhausted the tolerance toward virtually any criticism of offshore development. This situation was especially lamentable in the Caribbean, where the offshore approach was fraught with potential pitfalls. There is a need to scrutinize this path, reveal how it has performed elsewhere, and discern danger signs. (p. 259)

For some, offshore data processing represents regional maturity, a technological coming of age, a pathway to economic interdependence; for others, it is evidence of a deepening and more dangerous pattern of classic dependency. Regardless of perspective, it is clear that Caribbean nations, touted as the most digitalized region on Earth, are already effectively tied into global telecommunications networks. In industrial terms, this phenomenon represents a new phase in the metropolitan firm-Caribbean worker relationship and incorporation into the transnational labor community as low-level processors of data.

DATA PROCESSING

The redeployment of international work is a generic feature of what Castells (1993) calls "the new global economy of the information age"

---

[3]From time to time, this chapter will refer to the Caribbean as if it were one socioeconomic unit with uniform economic performance. That is, in fact, not the case. Each nation of the region experienced its specific economic history in the last 20 years. In the 1990s, however, on the whole it is fair to say that the region is undergoing a difficult period of economic adjustment.

(1993) or in Hepworth's (1990) phrasing, the new "geography of the information economy." By negating the importance of geographical proximity, that is, distance from the work site, new data management technology has facilitated a major shift in labor patterns worldwide, opening up ODE employment potential on a vast scale.

The advanced technologies, which ensure that certain categories of clerical and data management work are easily externalized, include technological application of EDI (Electronic Data Interchange), OCR (Optical Character Recognition), VDE (Voice Data Entry), BCS (Bar Code Systems), and IFP (Intelligent Forms Processing). Linkages to international data transmission networks encourage LDCs to focus their telecommunication investments in the following principal technologies and services: data entry, document imaging, computer-aided design/computer-aided manufacturing (CAD/CAM), Geographic Information Systems (GIS), telemarketing, and programming (INTEX USA, n.d.).[4] This allows them relatively easy market entry and competitive labor advantage. For foreign data-processing investors (FDPIs),[5] it allows for standardized quality, competitive costing, and locational flexibility—as long as control and entry aspects are regularized and labor is globally and competitively accessible.

Although software development, CAD/CAM, imaging, OCR, and other services are promoted in the Caribbean telecommunications export services sector, telemarketing, document imaging, and data processing appear to be the most strategically feasible to ensure near- and long-term viability (INTEX USA, n.d.). In the short term, demand is for low-end, highly manual functions, a sort of service station for information processing, or for telephone service operations, such as telemarketing. The latter, an option of some significance, is segmented into two classes of service: inbound (call receiving) and outbound (call making). Caribbean telemarketing has been used for direct-sale, large-volume client contact in short-time frameworks that call for minimal capital investments by the firm. Telemarketing's estimated $700 billion of sales worldwide by 1996 makes it an extremely attractive option for national accounts (INTEX USA, n.d.).

---

[4]INTEX USA is a consultancy firm specializing in telecommunications and development. The referenced text is from a study, "The U.S. Information Processing Market," done for the Jamaican government. The report is undated, but an interview with the principals revealed that it was done in 1991 or 1992.

[5]I use the term FDPI for "foreign data processing investor" rather than, for example, transnational or multinational enterprises (Castells, 1993) because it is unclear what type of firm does the bulk of investments in data processing. Although it is clear that many of these are classic transnational types, it remains to be seen whether all fall into this category.

Another target of development planners is document imaging, the electronic conversion of data into digital codes. It uses a technology similar to fax machines to convert, that is, digitalize, paper-based information to computer information. Two types of conversion practiced are archetypical of the industry: slow turnaround data entry (over three days) and fast turnaround (one to three days) data entry. They often, but not always, involve some sort of imaging. The former includes backfile conversion—digitalizing all existing information, including archival files, historic data, and personnel files of employees. The latter, recurring conversion, includes digitalizing the information that is currently used or generated on a daily basis. These include active files, credit card applications, airline tickets, car insurance forms, accounting receipts, and so on.

For large firms there are obvious advantages to document imaging: storage capacity, quick retrieval, safety, security, productivity, logistics, economy, and ecology. It allows for storage of very large amounts of data on optical disks (200,000 pages on a disk), efficient and selective retrieval of information, indexing of strategic information for easy access, routing and processing of vital information in minutes, maintenance of records safely for long periods, improved security and control of information, increased productivity in the workplace, and a reduction of costs associated with paper use, handling, and storage. However, conversion is labor-intensive, quality- and accuracy-dependent, and cost-sensitive. Hence, the nature of "the laborer" differs from the traditional concept. She must undergo some level of technological training and acquire a high level of data-processing efficiency (INTEX USA, n.d.).[6] At present, high-speed imaging technology is too costly for all but the largest data entry firms. One firm manager estimates that the cost of purchasing document imaging, sending, and receiving equipment for a Caribbean-U.S. linkup is US$85,000 (Howland, 1993).

Currently, the data entry subsector is the most developed of international information services in the Caribbean. Data entry operators "capture" data by keying in, scanning, or digitalizing information into computer-readable files. The information is either sent by bulk, using conventional transport (e.g., air), or by high-speed scanners from the customer's office to an offshore facility in a competitive labor market

---

[6]The "she" pronoun is used here, inasmuch as the overwhelming majority of data processors and data entry workers are female. The same is true for most export processing jobs, which demand patience, steadiness and dexterity of hand, long hours—and low pay. Indeed, the industry stereotype of the female worker is being reinforced by the backhanded "compliment" that "women are better at these jobs."

anywhere in the world. Documents will then be keyed in, either on-line, in interactive mode, or periodically transmitted in batch mode to the host computer. It involves both physical and automated work (e.g., optical character recognition and bar code readers), and its use is spreading across all sectors of industry that demand quality, efficiency, and speed in the processing and management of information.

A labor-intensive, low-to-moderate skills enterprise with inexpensive controls, outsourced data entry work has reportedly reduced by approximately 50% a typical data-processing budget in the United States (BIDC, 1994a). Total revenues for the information-processing market is estimated at about US$31 billion. Data entry is estimated to constitute about 20% of this market, or US$6.4 billion, and a growth rate of approximately 16% per year (INTEX USA, n.d.). So far, however, its Caribbean dividend has been principally in the employment sector, not in large inflows of foreign capital.

NIDL AND THE THIRD WORLD

Outsourcing of data entry and other clerical services is not a new phenomenon. Telemarketing, probably the earliest international telecommunications service work to enter the Caribbean commercial circuit, was first used in the 1960s. According to Data Entry Managers Association (DEMA) president, Norman Bodek,[7] data entry itself is an industry that's been "dying on the increase for the last twenty years" (quoted in Sachs, 1983, p. 74). Bodek estimated in 1983 that there were 600,000 people employed in data entry full time in the United States, and that "all offshore companies combined add up to little more than 3,000 jobs"(p. 74). Since Bodek's 1983 assessment, the transnational character of electronics labor has expanded around the world. More than 40,000 export jobs have now been targeted to LDCs in both fast and slow turnaround work (World Bank, 1994). In 1992 and 1993, between 2,500 to 4,000 data entry jobs were established in Jamaica alone.[8]

Clearly, what distinguishes this phase of development from Bodek's expectations are the intensity, pace, and rapid transnationalization of the industry. Over the last decade, these "quasi-

---

[7]DEMA, the Data Entry Managers Association, appears to have either dissolved or merged with another association, possibly the Association for Work Processors Improvement or Productivity Press.

[8]One must exercise caution in interpreting these figures. They are simply guidelines. There are almost as many numbers as there are writers. This is probably because of the volatility of the industry and the footloose nature of the FDPIs. The estimate here is taken as the most reliable range of sources.

markets" for information services have been increasingly "externalized" as part of the "informatisation of [the] economy . . . the redevelopment of information space through commoditisation, electronification and internationalisation" (Hepworth, 1990, p.122). An estimated 40 companies in the United States, England, and Japan have sent their labor-intensive clerical work to lowcost operations in places such as Ireland, Scotland, Sri Lanka, India, Taiwan, South Korea, the Philippines, Mexico, the People's Republic of China, and the Caribbean. Eastern Europe, Latin America, and Africa (e.g., Zimbabwe) are now considered potential data entry sites. Externalization of jobs occur principally because of the labor cost differentials between rich Organization for Economic Cooperation and Development (OECD) economies and those of the developing world (World Bank, 1994). Urban and rural labor pools, potential data processing sites in industrialized centers, and poor areas of newly industrialized countries have also been opened up, making the phenomenon truly global. For low-cost, data processing work, what technology makes possible, unemployment and underemployment in international labor make profitable, and transnational competition makes necessary. These industry characteristics have intensified the incentive to "outsource" work to "offshore" sites.

Transnational firms also use strategic locations as bases for expansion into related markets. Marketing logistics was a major reason for Texas Instrument's establishment of a facility in Bangalore, India, in 1986. According to company officials, the firm wanted "a presence in Asia which is expected to become a growing market for electronic equipment" (Shereff, 1989, p. 24). Firms focusing on data entry capitalize on cultural conditions to enhance economies of scope. SAZTEC International, which keystrokes information for IBM, telephone companies, law firms, and other business customers, has forgone because of high cost-to-quality ratios its Caribbean options to develop plants in the Philippines, where it now employs 900 people in data processing. It has also expanded to Ireland, enjoying favorable cost-to-quality ratios, and where cultural conditions are presumed to be unthreatening (Robert Eckholt, SAZTEC Vice President for Consulting Groups, telephone interview, June 1994).

Conceptually, the data entry phenomenon cannot be separated from the larger issue of transborder data flows and the process through which poor nations are incorporated into the world electronic community. Instead of utopian promises of a democratic, emergent "global knowledge grid" that will reduce gaps in vital knowledge worldwide (Dizard, 1990, p. 16), it is more plausible that international telecommunications networks will further divide a newly internationalized economy into economic zones in which LDCs serve as enclaves in information processing, rather than as equal members of a global knowledge community. Currently, economic conditions in many LDCs allow transnational enterprises to maximize their leverage with LDC governments, raising fears that these nations, or sectors of them, may become "electronic colonies." As competition picks up and the industry expands, the balance of negotiating power in ODE/ODP trades is expected to shift significantly away from soliciting governments and in favor of FDPIs, particularly as large firms progressively make instrumental use of production and locational flexibility to transcend local concentrations of trade union influence (Hepworth, 1990:123) and other types of regulatory regimes.

Economic survival is what inevitably draws most LDCs into the system. By staying out of telecommunications networks, nations face the real prospect of being left to fester on the edge of the world system (Castells, 1993; Payne & Sutton, 1993). Payne and Sutton's (1993) foreboding analysis is supported by a Caribbean prime minister, who warned his CARICOM colleagues in 1989 that the "Caribbean could be in danger of becoming a backwater, separated from the main current of human advance into the twenty first century" (p. 26). For these governments, employment (of almost any type) has been a strong enough political incentive to push them to secure unfavorable arrangements for national labor,[9] particularly when it comes with the veneer of international, "high-tech" sophistication. Thus, a new phase of "industrialization by invitation," centered on data entry work, the "bread and butter" of the telecommunication sector (Chang, JAMRO Group Director, Services Industries Division, personal communication, May, 28, 1993)[10] has become the panacea for economic survivability, particularly in the English-speaking Caribbean.

Socioeconomic and cultural conditions in the industrialized countries also affect the transnationalization of information work. This is because Third World demographics, so attractive to TNCs and FDPIs, also appear in rural and urban sectors of industrialized countries. Metropolitan labor pools remain underutilized because labor law and wage regulation, demographic, and cultural factors limit their usefulness to the TNCs and FDPIs. Cost is the most important consideration. In the United States, for example, the Office of Technology Assessment (OTA) estimates that, in general, off-shore

---

[9]It would only be fair to say that countries in the region, particularly Barbados and Jamaica, have set their sights on higher-end work in software development, programming, and systems design.

[10]As a representative of JAMPRO, Chang's view is obviously promotional rather than academic.

wages range from one fourth to one fifth of U.S. wages, with the lowest wages found in Asia (Ludlum, 1986). This ratio also holds true when comparing foreign wages with those of rural labor (Howland, 1991). The higher the U.S. minimum wage, the more foreign workers are priced into the market. Additionally, according to the U.S. Bureau of Labor Statistics, the number of 16- to 24-year-olds in the United States—the age group usually recruited for data claims processing—was projected to decline by about 10% and to remain depressed until 1997. Lower income minorities make up an increased percentage of this age group and traditionally have a large high school dropout rate. Minorities normally are not a target for determined and comprehensive economic development either by government or TNCs. It is unlikely then that minorities and women (whose shift from unpaid domestic labor to the workforce during the past 15 years is almost complete) will provide the reserve of educated workers waiting to be hired by data entry contractors (Shereff, 1989). As labor in industrialized nations becomes more costly and restricted through regulation, job definition, and demographics, LDC policy has been to expand the workforce by offsetting these tendencies.

Moreover, "pie in the sky" opportunities drive LDCs to liberalize initiatives to boost labor supply through narrowly specialized data entry training. Other initiatives are enacted to keep wages low and decrease workers' ability to organize. LDCs have implemented this restrictive policy by assigning ODE/ODP work into export-processing zones (EPZs, also called free trade zones—FTZs). EPZs are targeted to specific foreign enterprises, such as garments and low end manufacturing. The EPZs keep businesses "hermetically sealed" from commercial communities, that is, host country customs agencies, a development of post-World War II dependency relations.

Except in entrepôt cities, there were no free trade zones up to the end of the 1950s. The first FTZ was established in Ireland in 1959; in the Third World the first was Puerto Rico in 1962. The FTZ is now a standard arrangement for hosting foreign enterprises in Third World countries. The "informatization" of these zones through teleport[11] investments represents an accelerated requirement for service labor in these zones. For example, Jamaica's teleport development project, called Jamaica Digiport International (JDI), is expected to open a new era for data entry operators in the region (Chang, personal communication, May 28, 1993). JDI-type installations are ports of call for information processing. They provide "wayside enterprise" opportunities for

---

[11]Teleports are satellite Earth stations that bypass public telephone networks to provide companies with affordable, high-quality international telecommunications services (World Bank, 1994).

countries, or zones, peripheral to but networked into the main lines of international telecommunications circuits in the service of the metropolitan economies. This exotic veneer notwithstanding, teleport and traditional FTZ trade have several common features: They are engaged in the most basic aspects of production; the only local input these industries use is labor and, therefore, the only local cost is wages; capital investment is very small in TNC terms; technology transfer is extremely limited; employment is primarily female; and forward and backward linkages to the local economy is minimal (Girvan, n.d.; Willmore, 1993a, 1993b, 1993c).

DATA PROCESSING: THE CARIBBEAN

The ways in which work and technology have evolved in the Caribbean help explain the current relationship of transnational technology to the people. In 1870, a British company linked the English-speaking Caribbean to the rest of the world by means of a hookup between Jamaica and the new transatlantic cable (Cuthbert, 1987). Completed four years earlier, the cable was the defining communication technology of the time. Naturally, this link strengthened colonial and imperial ties between the colonies and the metropolis; therefore, neither internal infrastructure nor pan-Caribbean networks were developed or encouraged.

Functionally, telecommunications served as control mechanisms for the management of local enterprises and to facilitate social networking of European administrators and their small circle of local associates. Inadequately developed for local usage, telecommunications infrastructure never was engaged in the central themes of Caribbean life: self-rule, independence and nationalist movements, the emergence of national peasant classes, and the struggle to create profitable and equitable forms of labor. In fact, for the most part, telecommunications networks opposed these tendencies. Telecommunications largely served to rationalize Caribbean industry according to colonial economic and administrative objectives. Political fragmentation, monocultural economics, and parochialism were its legacy.

At the same time, telecommunications infrastructure reflects Caribbean reality: The islands are Europe's oldest industrial colonies in the Third World (Lewis, 1983); they remain the only viable geographical entity artificially created and, following the liquidation of their indigenous peoples, composed solely of immigrants purely for commercial ends (Allen, 1992). This is the core controversy, culturally and economically, of Caribbean life. It has directly influenced philosophies of development and the themes of Caribbean economics. The question is whether the adoption of electronic, export-oriented trade reinforces this history, or whether this trade will serve to liberate its people.

The aggressive pragmatism associated with data entry development comes, in part, from a loss of faith in "the experience of the 1970s and 1980s [which] shattered the idealism and naivete associated with Caribbean economic nationalism of the early post-independence period" (Ramsaran, 1993, p. 238). Indeed, the 1970s has been called "the lost decade of development." Across the region, in varying levels of intensity, a number of structural problems emerged: the oil crisis; worsening external debt; capital flight; declining standards of living; declining foreign exchange reserves; declining government reserves; savings and investments; declining nutritional standards; cuts in wages and salaries; reductions in the public servicework force; devaluation of currencies; and increased crime rates (Ramsaran, 1993).

Consequently, when American Airlines closed its data-processing facilities in Tulsa, OK in 1984 and moved 500 jobs to Barbados, the implications for labor captured attention both in the United States and the Caribbean. Observers believed that "after years of stagnation," hopeful signs were emerging "in this region of diverse cultures and dependent economies . . . opportunities for new partnerships were being presented as little known economic advantages became understood in the U.S." (Weiss, 1985, p. 8). Barbados, Jamaica, Dominican Republic, and the Eastern Caribbean states[12] launched campaigns actively promoting themselves as sources for U.S. data entry labor. Between 1982 and 1987, Jamaica attracted approximately 3,000 data entry jobs (Kuzela, 1987).

Predictions are that Jamaica can realize a 10,000 job increase over the next several years, and Barbados alone has attracted 11 data entry operations in the last decade (Howland, 1991). These estimations and tangible accomplishments are just the beginning. As early as the 1970s, Caribbean planners were aware that the data-processing industry would grow as a result of higher efficiency demands of metropolitan businesses in the information services sectors, and that international labor would be an important part of the informatization of the industry. It was evident to Caribbean data entry promoters and foreign data-processing investors that data-processing would bring a new era of industrial relations to the region, and they believed the region was well positioned to become full partners with FDPIs in expanding international trade (Bartlett, 1989).

---

[12]The OECS, the Organization of Eastern Caribbean States, consists of member states Antigua and Barbuda, Dominica, Grenada, Montserrat, St. Kitts and Nevis, St. Lucia, and St. Vincent, and the Grenadines.

OPTIMISM OVER TELECOMMUNICATIONS

Not surprisingly, massive investments in telecommunications technologies were made in the region to attract routine data entry jobs (Howland, 1991). With ITU encouragement, the region, like other parts of the Third World, "placed greater emphasis on telecommunications, and came to regard this sector as central to their development" (Demac & Morrison, 1989, p. 51). Thus, by 1990, the Caribbean had already moved, or was moved, to become an integral part of a global information network (Coles, 1990).

Driven by a mix of utopianism and pragmatism, Caribbean governments have undertaken sophisticated public relations campaigns to attract foreign data-processing firms. Working through international organizations such as the World Bank, United States Agency for International Development (USAID), and the Economic Commission for Latin American and the Caribbean (ECLAC), and consultancy firms such as INTEX USA and The Services Group, both of Arlington, VA, Caribbean nations have developed extensive blueprints for survival in the international data- processing sector. The consensus is that employment expansion will be an immediate payoff. Recent unemployment rates of 24% in Barbados, 30% in the Dominican Republic, and 25% in Jamaica, are distressingly high. To unemployed and underemployed nationals, data entry operators receive what are perceived to be relatively good wages (Sachs, 1983), and workers are easily captive thereby of the sophisticated information campaigns promoting the industry.

TRANSNATIONAL ENTERPRISES: ADVANTAGES

It was not lost on the more aggressive FDPIs that cultural proximity and advantages in labor conditions, limited regulation, politics, technology, culture, and the already vulnerable economic situation of Caribbean nations provided optimum opportunities for firms wishing to outsource. For transnational firms, the pluses (abundant, inexpensive labor; low staff turnover; high staff loyalty; underemployed, educated workers with English fluency in some countries; increases in use of idle computer resources, particularly night downtime, and government incentives) far outweigh the disadvantages. Concern about losing vital data and product control, privacy, security, cultural and language differences, political and economic instability, inadequate and inefficient telecommunications links, productivity shortfalls because of poor labor standards, opposition from labor groups, adverse laws and regulations, and the high cost of training and staff relocation are all disincentives for potential FDPIs (Anthes, 1991). Yet given that these barriers have not slowed industry growth in the region, it appears that for U.S. firms, the growing Caribbean economy and political atmosphere are positive prospects (Tierney, 1993).

FDPIs that Caribbean governments have been courting are service bureaus such as SAZTEC and foreign companies with large, daily internal data entry requirements, especially American Airlines[13] (USAID, 1988) and other corporate "end users."[14] Other major companies that use Caribbean labor through direct outsourcing and subsidiary operations or through service bureaus include: Sears & Roebuck; Dialog (financial services); BRS (financial services); Mead Data Central (publishing); McGraw Hill Publishing Company; NPD/Nielson, Inc. (research and publishing); Texas Instruments, Inc.; Digital, Inc. (computer software company); and BPD Business Data Services Ltd. Caribbean local firms sometimes collaborate with service bureaus in joint ventures.

JAMAICA

Jamaica is the largest of the 13 English-speaking Caribbean Community (Caricom) countries. With a population of 2 1/2 million, Jamaica has undergone one of the most ambitious data entry programs in the region. These enterprises have a choice of locating either inside or outside the publicly owned free trade zones. The government prefers the former because its strategy appears to be geared toward free trade zone-based, foreign, export-oriented, data processing enterprises. Three FTZs exist: the Kingston Free Zone, the Garmex Free Zone, and the Montego Bay Free Zone. The Montego Bay Free Zone is the location of the country's model data processing facility, Jamaica Digiport International (JDI). Information processing inspires heady optimism about its short-term employment potential. In a document titled, "Going for the Iceberg," the

---

[13]Companies such as American Airlines have found it profitable working under their corporate umbrella, AMR, to set up a subsidiary that conducts American Airlines work but also subcontracts to other data entry users. It first set up the subsidiary CDS (Caribbean Data Services), but this has changed relatively recently (1993) to DMS (Data Management Services) to accommodate more clients. The data management branch of the firm comes under the aegis of AMR's SABRE group.

[14]As of 1988, very few "end users" had actually set up data entry operations in the Caribbean. The trend may be changing but service bureaus still appear to be more eager to explore offshore opportunities.

Exporters of Information Services, a subgroup of the Jamaica Exporters'
Association, projected that by 1994, export earnings from the sector
would each $140 million, and employment would skyrocket to 18,000[15]
(JAMPRO Corporation, n.d.).

ECLAC economist Larry Willmore (1996), specialist in export
processing, describes the Jamaican economic situation that led to the
government's enthusiasm for the export processing (including garment
production) sector's potential:

> In the 1970s and early 1980s, Jamaica experienced a severe and
> prolonged economic recession. In fact, its gross domestic product
> (GDP) *contracted* at an average annual rate of 1.0% between 1970 and
> 1985, the worst performance by far of 23 countries listed in ECLAC's
> *Statistical Yearbook for Latin America and the Caribbean*. The recession
> caused unemployment rates to exceed 25% of the labour force. Lack of
> employment opportunities, combined with social tension, stimulated
> considerable emigration of skilled labour, including managerial talent.
> In 1986 a modest recovery began that continues to this day.
> Nonetheless, the economy remains highly dependent on the export of a
> few traditional commodities, all of which face depressed prices in
> world markets. (Willmore, 1994; emphasis in original)

The export-processing sector thus became one of the
government's highest priorities. It is expected to absorb an increasing
percentage of Jamaica's 900,000 total employees and brings in foreign
currency. Forty foreign and local data-processing firms, working in and
out of FTZs, account for 4,000 jobs (Chang, personal communication,
May 28, 1993). The foreign data entry firms employ just over 8% of total
export sector labor, with about 2,548 workers. From five, basically
domestic market, firms that existed in Jamaica 10 years ago, 26 (24
percent of all export processing firms) were operating in 1995.[16]

As the export-processing industry developed over the years, the
government offered new incentives packages (such as the 1956 Export
Industry Encouragement Act—EIEA,[17] to firms willing to specialize in

---

[15]Figures and predictions vary widely. It is unclear how these assessments are
made, but each reasonable assessment is taken at face value. It is an area that
begs research. Although no national denomination is given, it is presumed that
U.S. dollars is intended.

[16]Again, these figures are approximations due to the frequent changes in
companies' location and corporate arrangements an legal status.

[17]The Export Industry Encouragement Act (EIEA) dates to 1956 and allows
approved companies a holiday from taxes on profits and imported raw
materials, These incentives are granted for a maximum of 10 years, but it is a
simple matter for a company to change its name every five years and obtain, in

export markets—and declared that data entry services could not receive tax holidays and duty import privileges unless they located in FTZs. For firms operating outside the zones, new legislation would terminate the EIEA advantages in 1995. This will disadvantage local firms and likely force them to relocate to the Montego Bay Free Zone. The Garmex and Kingston FTZs are not competitively equipped (Willmore, 1993b).

JAMAICA DIGIPORT INTERNATIONAL

Earliest of the information processing firms to develop in Jamaica was Telmar Jamaica Ltd., jointly owned by Radio Jamaica and New Jersey-based Telmar Data Systems. In 1984, the firm installed a satellite dish that transmitted processed data back to the firm's U.S. clients (Cuthbert, 1987).

Currently, the standard in the data-processing industry and government FTZ approach is the Jamaica Digiport International (JDI), opened in 1989 to provide low-cost satellite communications for data-processing and telemarketing companies in the Montego Bay Free Zone. It is a joint venture between Telecommunications of Jamaica (30%), AT&T (35%), and Britain's Cable and Wireless (35%). Voice, text, and data handled by JDI are beamed to a satellite off the African coast and then to clients in the United States, Canada, and the United Kingdom. (Chang, personal communication, May, 28, 1993; JTURDC, n.d.). The JDI earth station in Montego Bay includes a 15-meter, 125 Watt, C-Band antenna. It offers its clients speeds between 9,600 and 1.544 megabits per second, international toll-free (800) numbers, credit card authorization for direct selling, and rates as low as US$0.24 per minute for calls to the United States (Willmore, 1993b).

The facility can be viewed as an example of electronic labor integration. Virtually every AT&T-type service is available from JDI at prices well below standard international rates (JAMPRO Corporation, n.d.). JDI provides technical and managerial support for the Jamaican data entry/data processing industry. Its license allows them to provide

---

effect, the incentive in perpetuity. The EIEA regime thus offers exporters in the customs territory nearly all the benefits associated with free trade zone status. The main difference is that EIEA exporters are not free from foreign exchange controls or from quantitative restrictions on imports. In recent years, with the abolition of exchange controls and the removal of import quotas, there is less distinction between the two regimes. In any case, the last EIEA incentives are supposed to end in 1995. (This information is extracted entirely from Willmore, 1993b, but it is corroborated through JAMPRO documents and personal interviews with JAMPRO officials.)

telecommunications services only to FTZ businesses. Operators outside the zone must use the more expensive Telecommunications of Jamaica (TOJ), the national telecommunications facility (Chang, personal communication, May 28, 1993). JDI optimistically expects that 30,000 Jamaicans will eventually be employed by the industry. By 1992, this figure was only 3,000 (Morris, 1992).[18]

ABSENCE OF JAMAICAN FIRMS IN FREE TRADE ZONES

Free trade zones were established for foreign, not domestic, investment. JDI, as a model of foreign-owned and -oriented export data processing, is expected to facilitate growth in that sector. Government policy to force local data processors into FTZs actually put them at a disadvantage with foreign-based enterprises. Competitive advantages for the foreigners become competitive disadvantages for the locals, leading to a "striking absence of Jamaican companies in free zones." The sole Jamaican data processing company in the Montego Bay Free Zone has the name "Bay Telemarketing Agency" because it could only obtain approval to operate there as a telemarketing agency. The company soon diversified into data processing, abandoning telemarketing along the way (Willmore, 1993b).

Not surprisingly, despite government overtures, some Jamaican data processors would prefer to work outside the FTZs. The manager of Bay Telemarketing announced that his company would relocate away from the zone or establish a satellite office at another location as long as the firm did not lose its export incentives. He claimed that the data-processing companies created too much demand for a small group of trained operators. Once trained by a company, they "are inevitably lost to other companies in the same building" (Willmore 1993b), or, in some cases, jump the industry totally (USAID, 1988).

Companies in the zones have a number of advantages. They can pay wages and other local expenses in Jamaican currency, but also maintain foreign currency accounts. They are exempt in perpetuity from taxes on profits, imports into the zone, and exports to other countries. Permanent customs offices are located in the zones, reducing red tape, and firms enjoy efficient granting of work permits for foreign nationals (JAMPRO, Corporation 1993).

Managers require that their staff enter the trade with at least functional literacy, preferably with a high school certificate and good typing skills. Although these are modest requirements, there is still a

---

[18]Again, the quoted figures are given face validity. They may differ from others provided elsewhere.

shortage of labor in the data-processing industry. Managers of data-processing firms allege that they could employ many more operators were they available and complain that data entry operators are hard to recruit. This means that the industry's growth prospects are constrained by supply rather than demand (Willmore, 1993b). Jamaica's low illiteracy rate (variously reported from 3.9% to 13%) and good, basic education makes it an ideal location for employing appropriately skilled workers. One alarming trend is the transfer of nurses to the data-processing industry, using their skills in medical transcription. The social consequences could be severe if this type of mobility is not adequately addressed.

Trainers with a good basic "commercial" education[19] can easily learn to operate programs for data entry and word processing. However, this pool of potential processors cannot fill the void, and there are several disincentives for industrial firms to train workers. New recruits normally require a firm's substantial investment. In general, "raw recruits" have such poor general skills that they require a minimum of three to four months of training. Although paid while in training, entrants are contracted at the legal minimum wage. Trained operators, always in demand, are under no constraint to stay with the employer who trained them.

A company that incurs the expense of training a worker must be able to recoup its investment; therefore, firms are not willing to risk investing much in peripatetic workers. As a result, "all firms invest too little in training new workers" (Willmore, 1993b, p. 15).

Data-processing firms are subject to the same labor laws as other Jamaican firms. Itemized operation costs for an entry-level keypunch operator in a nonfree zone environment include the basic minimum wage (as of 1991) of J$300 for a 40 hour week, equivalent to US$0.34 per hour (1991 exchange rate). This translates to US$1,835 annually. Apart from the basic wage, the employer must also pay payroll taxes totalling 11.5%. Fringe and discretionary benefits are assumed to cover 22% of the indirect labor wage bill and are negotiated based on accident insurance, health, uniform, laundry, retirement, and additional sick and maternity leave.

The total annual wage per employee, inclusive of fringe benefits, then is about US$2,140. Workers are assessed a 10% loss of US$214 to absorb absenteeism and training, which leaves them with a gross paycheck of US$1,926 per year. Jamaican labor laws require payment for vacation, maternity, and sick leave based on eligibility. Work in excess

---

[19]"Commercial education" refers to the basic training in business- centered subjects that students may pursue either in lieu of high school or after high school. The curriculum may vary from secretarial to accounting activities.

of eight hours per day and any work on Saturday must be paid at the rate of time-and-a-half. Double-time is paid for work on Sundays or public holidays. General managers primarily in FTZ employment are normally recruited from overseas on a temporary basis and earn an estimated US$14,323 annually, a small fraction of their U.S. counterparts. Their responsibilities include training of workers and supervisors, establishing systems of inventory and control, plant layout assistance, and training and development of a suitable understudy (JAMPRO Corporation, 1991b).

Keystroke and piece-rate systems are used in virtually all data-processing firms in Jamaica, making minimum wage only a wage floor. In a competitive industry in which quality and consistence are crucial, workers regularly producing below the established quotas are dismissed. The inverse is also true that, with such strong negative incentives, average wages are much higher than minimum wages. Positive incentives, such as productivity bonuses are also in place.

It is difficult to get industry managers to discuss salaries, but it appears that the average data entry operator in Jamaica earns considerably less than the average sewing machine operator (Willmore, 1993b:16) or hotel worker. Thus, when an information processor earns the same wage as a hotel worker, he or she takes home far less because the hotel worker pays no tax on tips, whereas "information processors pay tax on overtime and incentive bonuses" (Morris, 1992, p. 3).

Managers generally report wages to be actually in the range of J$500 to J$600 per week (US$0.56 to $0.68 per hour). This is still somewhat low for semiskilled employees, who have more formal education than ordinary factory or hotel workers. However, there are social status and non-pecuniary benefits associated with office and computer technology that make the data-processing profession compare favorably with similar level wage earners in other industries. For example, "data entry operators [in Jamaica] do work in a quiet, air-conditioned environment with flexible hours whereas garment factories are rarely air-conditioned and tend to be less pleasant" (Willmore, 1993b, p. 20). Table 4.1 provides one set of wage comparisons for the region and the United States.

BARBADOS

With a population of 253,000 and a literacy rate of over 98%, Barbados has a very ambitious offshore recruitment initiative in information technology services, including a sizeable commitment in data entry. A "friendly competitor" to CARICOM co-member Jamaica, Barbados'

Table 4.1. Comparative Positions of Selected Eastern Caribbean
Countries (hourly direct labor costs in US$).

	Hourly Wage for Data Entry (US$)	Hourly Wage for Secretary (US$)	Hourly Wage for Voice Operator (US$)	Cost for Dedicated kbs Int'l Service (Half circuit) US$
ST. LUCIA	1.10 (operator) 1.70-2.50 (suprvsr)	2.10-2.50	1.70	4118/mo
DOMINICA	0.80-1.00	2.50-3.50	2.25-3.56	4118/mo
GRENADA	1.26-2.10	2.42	3.05	4118/mo
ST. KITTS	1.40	2.25	1.80	4118/mo
ST. VINCENT	1.10-1.57	2.16	1.57	4118/mo
JAMAICA	.80-1.00	1.75	1.10	1850/mo
U.S	7.00-8.00	8.50-10.00	8.00-12.00	—

Data compiled from St. Lucia National Development Corporation (1994); ECIPS (1993-1994); 1992 TSG Field Studies; Cable & Wireless International Services Price List (1992)

Source: World Bank (1994).

Investment and Development Corporation (BIDC) promotional campaign aggressively touts the nation's world-class facilities, low operating cost, high accuracy, direct data transfer, confidentiality, and quick turnaround. As of 1989, the BIDC had subsidized 10 foreign data entry firms with a total of 1,200 employees (Howland, 1991). To date, a growing list of transnational firms take advantage of the Barbados program: R.R. Donnelley & Sons, in pre-press preparation; AMR (American Airlines), in airline ticket data entry; National Demographics, in warranty card processing; Southwest Data Base Co., in real estate transactions; Confederation Life Insurance (Canada), in insurance claims processing; Digital Imaging & Technology Services, in imaging of airway bills; and AJD Data Services (Canada), in imaging credit card applications. These and other companies are attracted to special incentives provided by the government for international service companies: full tax exemptions for U.S. foreign sales corporations, full tax exemption for "captive insurance companies," tax rate of only 2.5%

for international business companies, tax rate of only 2.5% on profits of information technology service companies, full and unrestricted repatriation of capital, profits, dividends, and cash grants for worker training.[20] Included in the package is facilitation of bureaucratic procedures and an official welcome from a government "committed to private enterprise and foreign investment" (BIDC, 1994b:9).

One of its biggest catches is American Airlines, which outsources about 900 pounds of used airline tickets every day by air to DMS (Data Management Services; formerly, Caribbean Data Services, as it is still called), an American Airlines subsidiary. The company estimates a savings of about US$3.5 million per year from the relocation (Beers, 1985); correspondingly, according to Howland (1991), Barbadian data entry operators earn an average of US$2.25 an hour, compared to a current minimum of $4.25 in the most rural areas of the United States.

DMS[21] or Caribbean Data Services (CDS) is the largest private-sector employer in Barbados. From an initial staff of 150 employees in 1983, CDS has grown to a staff of about 1,200. In addition to working for AMR, of which it is a division, CDS now generates about one third of its income from sales to outside customers on a contract basis (Nurse, 1992). The Barbados government boasts that CDS has gone on to become "the model that other international information services companies have adopted . . . and plans to incorporate scanning and other new technologies that will speed operations in the Barbados plant" (BIDC, 1994a).

Barbados offers large pools of unemployed and underemployed labor and has a reputation for productivity that is competitive with the United States. One U.S. firm operating in Barbados claimed it opened its branch there not only for its low labor cost but for the larger qualified labor pool (Howland, 1991). For example, when AMR moved data entry work from Tulsa, OK, not only were costs reduced by 50%, but quality and productivity surpassed that of the Tulsa facility (Nurse, 1992). In an interview conducted by Howland (1991), all offshore managers of U.S.-headquartered firms in Barbados concurred that foreign workers are as reliable and as accurate as the best rural workers in the United States. The illiteracy rate in Barbados is only 2%, much lower than that of the United States. All offshore firm interviews reiterated satisfaction with the skills and work ethic of the local workforce.

---

[20]This refers to an industrial training grant program. The Barbados Investment & Development Corporation will reimburse employers a maximum of 75% of wages paid to trainees during the first six months of operation. (maximum eight weeks per person; $35 per/week per trainee).

[21](see note 13). Many articles still refer to DMS as CDS. CDS is retained here for consistence. DMS is only used when convenient and necessary.

THE EASTERN CARIBBEAN STATES

Of all OECS (Organization of Eastern Caribbean States) member states, St. Lucia, Grenada, and Dominica have moved most aggressively into this sector by competing for data entry work in the international market, primarily in the United States. However, as data processing becomes more central to their economic diversification efforts, their vulnerability to itinerant FDPIs also becomes apparent.

St. Lucia, the largest OECS state with a land area of 616 square kilometers and a population of about 137,000, is representative of the industry. Although labor statistics are not very reliable, it is estimated that approximately 16,000 persons are employed in agriculture, 6,000 in manufacturing, 3,000 in hotels, and 24,000 in other services. Export processing firms (including garments)[22] employ more than 2,800 persons, amounting to half the jobs in manufacturing and 6% of all jobs in the economy (Willmore, 1993c).

St. Lucia entered export processing in 1968 with the establishment of U.S.-owned Manumatics Ltd. a plant that coils wire for transformers and fields of electrical motors (Willmore, 1993c), and moved into data processing in 1986, when Data Key International established a plant in Castries, the capital. In 1989, data processing represented only one firm, which employed 76 people, or 2.7% of the export processing sector. However, at the beginning of 1993, St. Lucia's 17 export processing plants were all doing information processing; three were U.S. owned and two were owned locally. The three FDPIs employed 88 persons or 3.1% of the export-processing sector. The U.S. plants were primarily data entry, whereas the locals attempted to work with programming, software development, general services, and training.

By mid-1994, only one U.S. firm (Atlantic International) remained, a reminder of the vulnerability to employer flexibility, competitive markets, and insecure partnerships with FDPIs. It was believed that two of the recent closings resulted because of difficulties in conducting marketing abroad, collecting payments due, and an inability to negotiate affordable international telecommunications services (World Bank, 1994). The two companies that service the local market still concentrate on programming, software services, and training, rather than data entry itself.

Atlantic International, the single international data-processing plant on St. Lucia, employed only 48 persons in mid-1994 (Williams, personal interview, 1994), a significant reduction from 88 employees one

---

[22]It must be remembered that "export processing" includes a wide variety of enterprises, but in the Caribbean, garments comprise the most significant proportion of the sector.

year earlier. Unlike Jamaica, the data processing operations in central Castries are conducted "in rather cramped offices with ceiling fans and no air conditioning" (Willmore, 1993c, p. 5). The National Development Corporation constructed a 20,000 square foot building designed specifically for data entry firms, but by mid-1994, the air-conditioned offices in the United Industrial Estates north of Castries had not attracted any tenants. Existing companies find rents in the new building beyond their means (Willmore, 1993c, p.5), and firms protest that the space cannot be easily partitioned. They also say the facility is poorly located relative to established telecommunications gateways and public transportation routes (World Bank, 1994).

The rapid turnover of international data-processing firms occurs in St. Lucia more than in other islands, "first, because of a shortage of trained or trainable data entry operators, and second because of the high cost of telecommunications" (Willmore, 1993c, p. 5). The government has addressed the former issue by sponsoring a 12-week Youth Skills Training Programme, which pays trainees a nominal stipend covering transportation and meals during the training period. The local telephone company has installed a fully digital earth station and fibre optic lines and allows for leased circuits and dedicated lines to the United States and Europe. All circuits to the United States terminate at Jacksonville, FL, where U.S. Telecom is the U.S. domestic service carrier. Yet the charges, equivalent to US$1.85 per minute for calls to the United States, are not competitive. For example, Jamaica's Digiport offers overseas calls for as low as US$0.22 per minute and "fast fax" connections that can bring prices of transmitting documents offshore to US$0.02 to US$0.03 per page. Furthermore, local firms face prohibitive price and access constraints when seeking to set up fast fax 56/64 kilobits-per-second circuits on a dialup basis (World Bank, 1994). A discount of 23% is in effect daily from 6 p.m. to 6 a.m. and all day Sunday. No other discounts are available to businessmen in St. Lucia (Willmore, 1993c).

There are other disadvantages. Uncompetitive prices exist for leased international circuits relative to alternative locations. In the case of Jamaica, monthly lease prices connecting the Caribbean investor to Intelsat satellites are less than US$1,850 for nonstop 56/64 kilobits per second links. The Eastern Caribbean, by contrast, quotes monthly lease prices of more than US$4,118 (World Bank, 1994). Furthermore, monopoly service agreements between OECS countries and their national communications providers (generally, a wholly owned or majority-owned subsidiary of Cable & Wireless) inhibit end user adoption of low-cost options (e.g., low cost satellite earth stations) for accessing satellite services directly (World Bank, 1994). These and problems of comparatively poorer communications infrastructure and costlier service charges restrict St. Lucia to the processing of articles for medical or legal journals and the like, which do not require fast turnaround times, whereas companies in Grenada and Dominica are able to service customers such as Federal Express, Bank of America, Xerox, and Blue Cross of California (Willmore, 1993c).

As a rule of thumb, Eastern Caribbean direct labor costs run from approximately 25% to 40% of those found in North America and Western Europe for a range of information processing services (World Bank, 1994). The St. Lucian government lists the wage rate as US$1 per hour for a trainee and from $1.10 to $1.50 for workers, depending on the level. A work inspector is listed at $1.65 to $2.50 per hour for supervisors, the highest category listed. However, Willmore reports that Atlantic International pays solely by the piece or keystroke. In these cases, the workers receive no holidays or vacation pay, although they are covered by the National Insurance Scheme (Willmore, 1993c). Wages are based on a standard 40-hour work week, excluding the cost of fringe benefits, which are a legal minimum of 15.4% (5% employer's contribution to the National Insurance Scheme, 13 paid holidays, and fourteen working days annual vacation). The government offers up to fifteen years of tax holiday, duty free entry and exit of equipment and other materials needed for operation, "no double taxation," and full repatriation of capital.

One OECS country, Grenada, successfully wooed DataLogic Limited (U.S.) in 1989. DataLogic (Grenada) now employs 150 data entry operators and has another 30 people in training. Apparently the firm has shown enough confidence to establish a teleport Earth station, which has "led to the creation of more than 200 jobs" (World Bank, 1994, p.II-4). Workers at the facility receive 100,000 images of waybills daily for processing from a key North American client, Federal Express. This one customer could ultimately create as many as 2,000 jobs offshore given its 1.2 million daily waybill volume (World Bank, 1994).

In 1992, the parent DataLogic acquired Offshore Keyboarding Corporation, previously owned by Cable and Wireless (the local telephone company) in Dominica and renamed it DataLogic (Dominica) Limited. Employment there since has grown from 85 to 115 (Willmore, 1993c). Success in both places is credited to the fact that the investors installed a satellite dish, enabling them to send and receive images of documents on their own equipment, rather than having to ship them or pay high fees to the local telephone company, a foreign-owned monopoly.

However, Dominica's small size makes it sensitive to sudden changes in employment. One foreign-owned facility (Digital Imaging), employing 100 workers, shut down in Dominica within the past 14 months, primarily because of financing problems. The company, however, continued to maintain workforces in Grenada and Barbados. A Dominican-owned venture in digitalizing maps and engineering documents foundered after training 30 people because of nonperformance by an oversees partner, which was to take responsibility for international marketing and promotion (World Bank, 1994).

DOMINICAN REPUBLIC

Willmore provides in the Dominican Republic case study a clear picture of how export processing zone (EPZ) policy is intended to function. EPZ and tourism have been the only dynamic sectors of the Dominican economy in the past decade. In the 1982-1991 period, gross domestic product grew by an average of only 1.2% a year, exports of goods from the customs territory fell in both volume and (U.S. dollar) value, and manufacturing production outside the free trade zone decreased. Over the same period, exports of FTZs grew 23% a year to more than US$1 billion, and foreign exchange earnings of the zones increased by an average of 17% a year to US$249 million (nominal dollars, not adjusted for inflation). It is notable, however, that FTZ employment increased sevenfold, equivalent to a compound growth of almost 25% a year, over this nine-year period (Willmore, 1993a).

Gulf & Western Corporation, a large sugar producer, which at the time owned 8% of all of the arable land in the country, opened the first FTZ in the Dominican Republic at LaRomana in 1969 (Willmore, 1993a). There are now 23 industrial FTZs in operation, with two more ready for operation and another under active development, in locations throughout the country. Eleven of the 25 zones are privately owned, 13 are government-owned, through the Corporacion de Formento Industrial, and one is of mixed private/public ownership. In the late 1980s, CODETEL (a subsidiary of GTE) in the Dominican Republic established a special program responding to FTZ telecommunication needs (Johnson, 1989). It is therefore surprising that, given its level of FTZ development, the Dominican Republic is not a major contender in the data-processing business (Howland, 1993) and remains one of the least developed, potentially competitive nations.

The catalyst for the Dominican data-processing industry might be American Airlines and its AMR data entry contractors. In 1987, the American Airlines subsidiary, Caribbean Data Services (CDS), set up a sister company to their Barbados plant in the San Isidro Free Zone. The company now employs 600 keyboard operators who handle 10% of reservations for American Airlines (Willmore, 1993a). It also handles the

American Frequent Flier program (AAdvantage) and other customer service activities, including all names, addresses, and complaints, and carries out special information-based services for outside clients. For example, the "Dominican office is linked on-line to an insurance customer's computer, via satellite, to enter coding on claims, the first step before an analyst processes them" (Shereff, 1989, p.26).

Shereff wrote that the strategy seems to take advantage of the growing corporate trend to contract out data entry work by seeking long-term agreements to handle other companies' in-house work. Three years after the CDS venture, Telepuente San Isidro was built in the expectation that improved communications with the United States would stimulate development of the information-processing industry. To date, however, Caribbean Data Services remains the only data-processing company in a free trade zone, and it continues to purchase international telephone service from the local telephone company rather than from Telepuerto San Isidro (Willmore, 1993a).

CDS hires only secondary school graduates with typing skills and is very selective in choosing those who are allowed to graduate from the four-week basic skills course they provide. The firm is even more restrictive in how many they select from this group to enroll in a further four weeks of on-the-job training in data entry. This is in sharp contrast with other export-processing enterprises that are mere assembly operations, characterized by repetitive tasks that require fewer skills (Willmore, 1993a). One might speculate that language plays a role in this division of data entry labor between the English-speaking and Spanish-speaking nations, although it is not clear that the distinction is the only factor. In any case, the Dominican Republic's EPZ strategy remains in sectors in which the overwhelming majority of workers have little education and receive no more than two or three months of on-the-job training (Willmore, 1993a), a strategy that has not attracted significant FDPI attention.

CONCLUSION

The foregoing overview provides a descriptive, critical outline of the Caribbean offshore data entry industry. The industry's arrival dovetails with the currently prevailing economic "liberalization" and "deregulation" policy, which is expected to "jump-start" regional economies. This radical move away from the "self-development" nationalism of the 1970s brings the region full circle, back to the (not too successful) "industrialization by invitation" strategy of the 1950s. Four characteristics (competition, speed of transmission, access tools, and interoperability) pace the use of telecommunications as a business resource (Keen & Cummins, 1994) and have resuscitated traditional patterns in the international division of labor. At the immediate, pragmatic level, this promises to be an economic windfall for participating LDCs and a great opportunity for FDPIs. However, despite the euphoria associated with it, this new level of work does not transcend the historical characteristics of the plantation mode of production to which the region is heir. The structure of labor and its formation through export-processing zones closely resembles work in other traditionally exploited industries such as the garment industry, although, in contrast, there may be prospects for development in professions associated with computing. To date, exotic telecommunication and information technologies' primary role in Caribbean circles has been to bring service work to cheap labor. Caribbean, English-speaking labor provides an ironic mooring for such activities, and the region's economic history as a provider of raw materials is the best cautionary note to the industry's development.

Obviously, the redeployment of labor to the region is an inevitable aspect of the global informatization process in which capital, production, management, markets, labor, information, and technology are organized across national boundaries (Castells, 1993). However, there are different views of how nations come to participate or are brought into the global network. An obvious starting point is the aforementioned international (spatial) division of labor. According to Howland (1991), "in the spatial division of labor model, international labor costs drive location decision, with management capabilities and industrial structure influencing the scope of locations under consideration" (p. 53). Her disagreement with this thesis comes from the perspective that technology plays a more critical role in the international distribution of electronic work. She argues that "a more appropriate model is one where technology plays the central role and influences location decisions by either freeing firms to minimize labor costs or, on the other hand, rendering labor irrelevant altogether. As the technology becomes more sophisticated the spatial division of labor model becomes less relevant" (p. 53). This view is countered by Pearson and Mitter's (1993) claim that "the use of offshore operations reduces the cost of data entry by a factor of two or more" and makes possible "the conversion of a whole series of records and information to machine-readable forms which would not be undertaken in the absence of this cheap option" (p. 57).

The structure, cost, location, and composition of the global labor force are compelling arguments for a sociological or political-economic interpretation of the international data entry phenomenon. Worldwide, the industry is expanding through developments of teleports in free trade zones, extremely competitive labor markets, global sourcing, and predominant female labor. Among these factors, the most dramatic is the overwhelming presence of women in the international data entry workforce.

This phenomenon lends special credence to the "dual division of labor" theory. According to Caribbean economist Norman Girvan (n.d.), the labor market, both nationally and internationally, is divided into dual segments. One is relatively skilled, enjoys relatively high wages, is highly unionized, and has substantial opportunities for advancement, upgrading of skills, and increased wages over time. The other segment is characterized by low wages, unskilled or semiskilled labor, with few opportunities for upskilling or for wage increases and is not usually unionized.

A major distinction of this segment of the export-processing labor force is its primarily female composition (Girvan, n.d.). McCormick (1992) documented that although women are complimented for being more compliant than male counterparts (see also Barry, Preusch, & Wood, 1984), and are more conscientious, they cooperate out of fear of losing their jobs (McCormick, 1992; Posthuma, 1987). Girvan (n.d.) presented additional documentation from researcher Victoria Durrant-Gonzales, who concluded from interviews conducted with export processors that these firms "hold and exploit cross cultural views that women have a natural proclivity for work that is tedious and monotonous" (p. 7). For example, a government official whose job is to attract U. S. business to the Bermuda Industrial Park, volunteered that "a man just won't stay in this tedious kind of work, he would walk out in a couple of hours" (Girvan, n.d., pp. 7, 8). These rather brief extracts are sufficient to make the point:

> Whilst the export of software and computer services by LDCs, as well as the growing use of computers in the domestic economy, have created high-skilled employment, mainly for men, the growth of offshore data-entry work and the computerization of data processing in these countries have created jobs mainly for women. (Pearson & Mitter, 1993, p. 53)

This bias toward female recruitment is shared by the FDPI, as well as by governments' promotional agencies, which attribute their hiring practices to women's "special attributes" in this sector.

Demographic factors also contribute to the concentration of women in these jobs. In Jamaica, as well as Barbados, women have become the primary wage earners in a high percentage of households. One can argue that it is not technology, per se, that influences this aspect of data entry work, but sociocultural factors. Regardless, "it is indisputable that low-skilled information processing is predominantly female and seems likely to remain so" (Pearson & Mitter, 1993, p. 53) while the industry exists.

Data entry work is highly supervised. In the Caribbean, where close foreign supervision is anathema, this creates sociocultural problems. Often, a minimum quota is used to pace workers, who may be fired if the quota is not met. According to McCormick, some firms constantly, and automatically, monitor their workers. Their word-processing terminals, "which are programmed to record speed error rates," act as "electronic supervisors" (McCormick, 1992, p. 20) and serve to intensify the pressure and pace of work. Some firms tabulate, extract, and post this information daily, weekly, or monthly to encourage greater competition and speed, in effect publicizing workers' adjusted pay scales and bonuses.

Just as striking as the female composition of the data entry workforce is its remarkable structural resemblance to the garment industry—which tells a lot about its restricted possibilities as a catalyst for national development. The garment industry too is primarily female and, in general, is sourced from the same international vendors. Clothing manufacturers based in Western Europe and North America have shifted the labor intensive segments of the industry, mainly assembly, to LDCs. This "redeployment" of labor is toward the repetitive, tedious, and routinized aspects of the work, and the existence of low-cost labor is essential to its profitability and viability. It is based on assembly (putting together of parts composed elsewhere), and design and higher skilled, better paid labor is done elsewhere. Unlike their LDC counterparts, metropolitan-based, higher end workers are unionized or at least organized and politically supported by lobbyists. The garment industry is a fixture of LDC free trade zones and is often braced by "liberalized" laws and regulations designed almost exclusively for export markets and foreign clients. These firms are notorious, itinerant seekers of cheap labor. They do not require a high capital investment and, therefore, can relocate overnight without much loss. Finally, backward and forward linkages to the local economy are minimal in the garment industry and less so in the data entry market. They are not integrated into the national commercial community. They require no material content from the host location, nor do their products spawn, support, or supply local industries.

One underestimated, long-term impact of the industry is likely to be the health problems associated with intensive use of VDU/VDT (video display units/terminals). Potential hazards in this area are likely to be a result of the Caribbean governments' "underregulation" industrial policy. So far, this seems not to be an issue, possibly because the industry does not have a long history, and no regional studies have been carried out to determine its effects—although they have been determined elsewhere.

This "head in the sand" policy may serve neither the industry nor its workers in the long run. Industry leaders should note that in industrialized countries, "where the dissemination of new technology is more thoroughly monitored by government agencies and labor organizations, the existence of potential health risks arising from the intensive use of computers is well documented" (Pearson & Mitter, 1993, p. 62). The quoted authors list five computer-based, information-processing health hazards: musculo-skeletal disorders, deterioration of visual capacity and related problems, stress and fatigue, skin complaints, and reproductive hazards. These conditions are attributed to poor ergonomic design of workstations, radiation emission from VDUs, static electricity and chemical emissions, and overuse of eye and muscles without break or rest. Research in Malaysia and Brazil document the existence of many of these data entry-related maladies. Because the labor force is composed primarily of women, research that reports a higher rate of abnormal pregnancies and deliveries by women who spent long, uninterrupted periods in front of a VDU (Pearson & Mitter, 1993) should be taken with utmost seriousness.

Although it is unlikely that data entry will "die on the increase" for several more years, the status of the industry prompts careful reflection and analysis. The most severe critics argue that technological innovations will ultimately remove completely the demand for cheap labor for data entry work. Howland's (1991) view is that "within the next decade, breakthroughs in scanning will permit the mechanical entry of handwritten documents, displacing manual data entry altogether" (p. 55). Micossi (1986) predicted that offshore data entry will die in the next 20 years. For the insurance industry, which was criticized for taking its claims processing offshore in the 1980s, one person's view is that "the great rush offshore to cut administrative costs have ended" (Iannello, 1992, p. 24). It is likely that there will be a shift away from the Caribbean to more cost-depressed areas in Asia. Concerning the disappearance of the industry, the handwriting is already on the wall: The industry will change profoundly in the near future. Charles Crichton, who directed the Data Conversion and Transmission Task Force, a sub-committee of President Reagan's U.S. Business Committee on Jamaica, although in opposition to the "disappearance thesis," cautioned that although offshore data entry may increase, it is not possible to predict the extent (or nature) of expansion (cited in Sachs, 1983).

Optical scanners, which electronically scan and digitalize printed pages into computer readable codes, voice data systems (VDS), which allow vocalizations to be received and coded, and significant improvements in other systems (see earlier discussion on data processing) will all affect the direction of the industry and the composition of its labor. Because these advancements in technology will require higher levels of training and investment, either the host governments or the FDPIs, or both, will have to undertake the levels of new capital investments required to keep the industry going. Defaulting on this option will certainly cede the industry to non-Caribbean competitors, who are already making significant inroads. In addition, faster telecomputing systems with greater online capacity will negate time zone differences and actually create an advantage for round-the-clock "day-time" workers from Asia as night falls on the Western hemisphere. However, Howland (1991) sees a boomerang to North American metropolitan labor. She argues that "when scanning equipment requires a highly trained labor force to program scanners for new jobs and maintain the equipment, offshore sites will return full circle to the urban areas of industrial countries" (Howland, p. 55).

The Caribbean industry is also likely to feel the effects of political pressure by groups lobbying on behalf of U.S. clerical workers. The U.S. Office of Technology Assessment has already noted that offshore data entry could gravely affect levels of clerical employment in the U.S. If predictions hold true that the new information economy will provide fewer new U.S. jobs than forecasted, labor unions wary of clerical work relocation can be expected to lobby against footloose transnationals. Redeployments to the urban and rural United States, Puerto Rico, and NAFTA communities in Mexico need also be considered.

Some industry analysts also believe that cost and security, rising telecommunications costs, the declining dollar cutting into the savings of lower wages abroad, and corporate hesitation for cultural reasons are still serious resistance factors. The last is important because the corporate customer base is not likely to shift significantly, and end-users have already demonstrated reluctance to move their data entry work offshore (USAID, 1988).

All these issues must be given serious consideration if the Caribbean industry is to be viewed in terms of long-term strategy. Caribbean telecommunications policymakers, currently engaged in regional integration and teledevelopment, have but two options: a continuation of the trend of the 1980s of closer integration, territory by territory, into the U. S. system; or the uncharted option of regional integration across the entire Caribbean (Payne & Sutton, 1993). The data entry industry, considered part of telecommunications policy, is *ipso facto* of the first trend. The dilemma is that data entry development itself contradicts the integration thesis. Any restriction in the industry will increase competition within the Caribbean community. As other Caribbean English-speaking nations get on the bandwagon, the prevailing temperament of cooperation and cordial competition (Chang, personal communication, May 28, 1993) is likely to change, as islands look out for their own interests.

To their credit, Caribbean states have created an impressive complex of national and regional organizations and telecommunications interest groups. The Caribbean Community (CARICOM) has taken a more active role in regional affairs. CANTO (Caribbean Association of National Telecommunication Associations), the CTU (Caribbean Telecommunications Union), CTC (Caribbean Telecommunications Council), and the TRINCOM (Trinidad and Tobago Communications), are specifically mandated to oversee and study the development and impact of international telecommunications infrastructure on Caribbean life. One of the CTU's major projects is a study of the region's technological options, including fiber optics, microwave facilities, and the feasibility of a Caribbean satellite system (Demac & Morrison, 1989), all of which will affect data entry enterprises.

Yet, except in the area of training, telecommunications development models have de-emphasized the worker. The tilt has been toward the study of the investment process, telecommunications products, and management structure. More attention has been given to contract employers and consulting firms and less to on-line data processors, the worker, collective bargaining, and labor-centered issues in general. The bias, however, is evident. More emphasis on tough bargaining will result in less FDPI presence. While it lasts, the data entry industry does provide some degree of employment and foreign exchange relief to Caribbean nations that take advantage of its opportunities. However, if one were to focus on the single female breadwinner of a Caribbean household, "under conditions of high unemployment and high inflation...simply having work becomes more important than organizing around the conditions of work" (Posthuma, 1987, p. 49; also cited in McCormick, 1992, p. 21). In the final analysis, the responsibility for and consequences of the data-processing investment strategy for development rest with the Caribbean leadership.

REFERENCES

Allen, E. A. (1992, May 3). Science & technology—engine of progress. *The Sunday Gleaner*, p. 12A.
Anthes, G. (1991). U.S. firms go offshore for cheap DP. *Computerworld*, 25(34), 1, 59, 60.

Barbados Investment & Development Corporation, Inc. (BDIC). (1994a, April). *Barbados: Gateway to international businesses* (promotional brochure). New York: Author.

Barbados Investment & Development Corporation, Inc. (BICD). (1994b, April). *Barbados: The ideal location for your business* (promotional brochure). New York: Author.

Barry, T., Preusch, D., & Wood, B. (1984). *The other side of paradise: Foreign control in the Caribbean*. New York: Grove Press.

Bartlett, R. (1989). Informing the Caribbean. In *Proceedings of Trincom-89: Caribbean Media and Telecommunications in the Information Age* (pp. 21-24). Port of Spain, Trinidad and Tobago: Inprint Caribbean Ltd.

Beers, D. (1985). Offshore offices: Corporate paperwork sneaks out of the country. *Working Woman, 10*(9), 55-57.

Castells, M. (1993). The informational economy and the new international Division of Labor. In M. Carnoy, M. Castells, S. Cohen, & F.H. Cardoso (Eds.), *The new global economy in the information age: Reflections on our changing world* (pp. 15-43). University Park: The Pennsylvania State University Press.

Coles, T. (1990). Accessing the market. In *Proceedings of Trincom-90: The role of the Caribbean in the global information village* (pp. 114-117). Port of Spain, Trinidad and Tobago: Inprint Caribbean Ltd.

Cuthbert, M. (1987). Communications technology, culture and development: Caribbean perspectives. In R. Hinds (Ed.), *Communications for development: Claiming a common ground, a Caribbean perspective* (pp. 27-37). Bridgetown, Barbados: Caribbean Conference of Churches.

Demac, D. A., & Morrison, R. J. (1989). US-Caribbean telecommunications: Making great strides in development. *Telecommunications Policy, 13*(1), 51-58.

Dizard, W. (1990). *The coming information age: An overview of technology, economics, and politics* (2nd ed.). New York: Longman.

Girvan, N. (n.d.). The international division of labor and free tradezones. In *Symposium on Free Trade Zones*. Kingston, Jamaica. Available from: Joint Trade Unions Research Development Center, Kingston, Jamaica.

Hepworth, M. (1990). *Geography of the information economy*. New York: Guilford Press.

Howland, M. (1991). *Rural computer services in an international economy* (Research rep. to the Ford Foundation). College Park: University of Maryland, Department of Urban Studies and Planning.

Howland, M. (1993). Technological change and the spatial restructuring of data entry and processing services. *Technological Forecasting and Social Change, 43*, 185-196.

Iannello, L. (1992). Waiting for the tide to turn. *Insurance Review*, *53*(9), 24-26.

International Technology Exchange Group (INTEX USA). (n.d.). *Overview of the U.S. information market* (Market analysis prepared for the government of Jamaica). Alexandria, VA: Author.

JAMPRO Corporation. (1991, March). *Data processing direct wage analysis: Basewage and fringe analysis. Notes/Assumptions associated with data entry operations in a non-free zone environment* (Report for the JAMPRO Corporation Feasibility Analysis Unit). Kingston Jamaica: Author.

JAMPRO Corporation. (1993, May). *The Jamaican advantage: An economic alternative*. Kingston, Jamaica: Author.

JAMPRO Corporation. (n.d.). *Overview of the Jamaican information processing sector*. Kingston, Jamaica: Author.

Johnson, T. (1989, June 19). Caribbean basin becomes bit telecommunications market. *Business America*, pp. 16-18.

Joint Trade Unions Research Development Centre (JTURDC). (n.d.). *Jamaica Digiport International Report for the JTURDC Education Department*. Kingston, Jamaica: Author.

Keen, P., & Cummins, J. (1994). *Networks in action*. Belmont, CA: Wadsworth.

Kuzela, L. (1987). New Jamaican teleport to serve U.S. business. *Industry Week*, *234*, 64-65.

Lewis, G. (1983). *Main currents in Caribbean thought; The historical evolution of Caribbean society in its ideological aspects, 1492-1900*. Kingston, Jamaica: Heinemann Educational Books (Caribbean).

Ludlum, D. (1986). Offshore data entry pays off. *Computerworld*, *20*(23), 103, 112.

Maingot, A. (1993). The offshore Caribbean. In A. Payne & P. Sutton (Eds.), *Modern Caribbean politics* (pp. 259-276). Baltimore: The Johns Hopkins University Press.

McCormick, P. (1992). The Caribbean: A paradise for offshore activities. In *Proceedings of the 17th Annual Third World Conference* ([draft version]. Chicago: TWCF Publications.

Micossi, A. (1986). Data entry: Better in Barbados? *Computer Decisions*, *26*, 20-21.

Morris, M. (1992, August 31). Digiport—a window to the world. *The Daily Gleaner*, p. 3.

Nurse, L. (1992). Barbados: An action center for information services. *Telemarketing Magazine*, *11*(4), 78-80.

Payne, A., & Sutton, P. (1993). Introduction: The contours of modern Caribbean politics. In A. Payne & P. Sutton (Eds.), *Modern Caribbean politics* (pp. 1-27). Baltimore: The Johns Hopkins University Press.

Pearson, R., & Mitter, S. (1993). Employment and working conditions of low-skilled information-processing workers in less developed countries. *International Labor Review, 132*(1), 49-64.

Posthuma, A. (1987). *The internationalization of clerical work: A study of offshore office services in the Caribbean.* Brighton, UK: University of Sussex Science Policy Research Unit.

Ramsaran, R. (1993). Domestic policy, the external environment, and the economic crisis in the Caribbean. In A. Payne & P. Sutton (Eds.), *Modern Caribbean politics* (pp. 238-258). Baltimore: The Johns Hopkins University Press.

Sachs, R. (1983). Should you send your input overseas? *Office Administration and Automation, 44*(3), 70-74.

The Services Group, Inc. (TSG). (n.d.). *Offshore information industry development services* (fact sheet). Arlington VA: Author.

Shereff, R. (1989). Creative clerical solutions: Service firms open up global satellite offices. *Management Review, 78*(8), 24-27.

Tierney, R. (1993). Caribbean, see? *World Trade, 6*(8), 78-84.

U.S. Agency for International Development (USAID). (1988). *The potential for data entry operations in the Caribbean.* Washington, DC: The Office of the Private Sector Coordinator of USAID.

Weiss, J. (1985). The Caribbean nears liftoff. *Managing, 1*(1), 8-11.

Willmore, L. (1993a, September). *Export processing in the Dominican Republic: Ownership, linkages and transfer of technology* (General WP/93/15). New York: United Nations Economic Commission for Latin America and the Caribbean. Caribbean Development and Cooperation Committee.

Willmore, L. (1993b, September). *Export processing in Jamaica: Ownership, linkages and transfer of technology* (General WP/93/13). New York: United Nations Economic Commission for Latin America and the Caribbean. Caribbean Development and Cooperation Committee.

Willmore, L. (1993c, September). *Export processing in St. Lucia: Ownership, linkages and transfer of technology* (General WP/93/14). New York: United Nations Economic Commission for Latin America and the Caribbean. Caribbean Development and Cooperation Committee.

Willmore, L. (1994, April). Export processing in Jamaica. In *CEPAL Review* (No. 52, pp. 1-23). New York: United Nations Economic Commission For Latin America and the Caribbean. Caribbean Development and Cooperation Committee.

World Bank. (1994). *Opening the information industry marketplace: Opportunities for Eastern Caribbean exports.* Arlington VA: The Services Group (TSG).

5

New Chips in Old Skins: Work and Labor in Silicon Valley

Lenny Siegel

California's "Silicon Valley," roughly covering the metropolitan portion of Santa Clara County, is the world's number one address in high technology. It is not only the most significant concentration of high-tech industry and talent, but the term *Silicon Valley* itself has become synonymous with a new way of doing business, based on continuous innovation and industrial flexibility. President Clinton and Vice President Gore, among others, have embraced the Silicon Valley experience as the model for U.S. industrial growth in the new millennium.

Silicon Valley's products are truly impressive and the performance of its leading companies remarkable, but the "model" needs changes if it is to serve well the country or the world. Silicon Valley is home to a large, well-paid professional workforce, but the Valley's high-tech companies directly and indirectly employ a vast number of poorly paid workers locally, nationally, and internationally.

Most are women, and most are immigrants or non-white minorities. Thus, the social structure of the "industry of the future" is polarized along class, racial, and gender lines.

In fact, many of the trends evident in Silicon Valley are now discernible in southern California as in other high tech centers across the nation. Nowhere, however, is high-tech so concentrated as in Silicon Valley, and the stratification of its workforce is a signal of what is to come elsewhere.

AN ENVIABLE ECONOMIC RECORD

Seven of the 21 California-based firms on the Fortune 500 are headquartered in Silicon Valley, which is also home to five of the eight largest U.S. merchant semiconductor producers ("The Fortune 500," 1995). Twelve of the top 33 information technology companies in North America, other major high-tech firms such as IBM and Lockheed-Martin Missiles and Space, and, overall, about 1,500 of the 2,500 largest U.S. electronics businesses are based in the Valley (see Table 5.1).

Equally important, while outside giants, such as IBM and Digital Equipment struggle to adapt to the rapidly evolving high-tech marketplace, Valley-based firms such as Intel, Sun Microsystems, Hewlett-Packard, and Conner Peripherals are growing and profiting. Paced by Intel's $1.1 billion in net income (as of 1992), the profits of the top 25 publicly-owned Valley firms jumped 46% from $2.67 billion in 1991 to $3.89 billion in 1992, despite the recession and the pressure of Japanese and other foreign competition. Revenues of the top 100 firms in Silicon Valley have nearly doubled since 1989 to more than $106 billion, and profits jumped from about $3 billion in 1991 to $7.564 billion in 1994 ("Silicon Valley 150," 1994).

PHYSICS DRIVES THE STRUCTURE

The manufacturers of silicon-based integrated circuits, or chips, are no longer the largest employers in the Valley, but the physics of the chip remains largely responsible for the Valley's unusual dynamism.[1] A few

---

[1]The industrial sectors included here in high-tech electronics manufacturing are the following:

SIC Code	Industry	# of reporting units
357	Office and Computing Machines	46
366	Communications Equipment	35

Table. 5.1. The Valley's Top High-Tech Firms (Locally Based, Publically Owned).

Company	Headquarters	1994 Sales ($)	1994 Profits ($)	Top exec. comp.($)
Hewlett Packard	Palo Alto	24,991,000,000	1,599,000,000	1,699,974
Intel	Santa Clara	11,521,000,000	2,288,000,000	7,590,236
Apple Computer	Cupertino	9,188,700,000	310,200,000	not listed
Sun Microsystems	Mountain View	4,689,900,000	195,800,000	not listed
Seagate Technology	Scotts Valley	3,500,100,000	225,100,000	5,608,000
Conner Peripherals	San Jose	2,365,200,000	109,700,000	4,266,566
Advanced Micro Devices	Sunnyvale	2,134,700,000	341,100,000`	8,887,210
Quantum	Milpitas	2,131,100,000	2,700,000	not listed
Tandem Computers	Cupertino	2,180,000,000	170,200,000	not listed
Oracle Systems	Redwood City	2,001,100,000	283,700,000	4,993, 289
Applied Materials	Santa Clara	1,659,800,000	220,700,000	5,621,289
Amdahl	Sunnyvale	1,638,600,000	74,800,000	1,809,468
Varian Associates	Palo Alto	1,552,500,000	79,400,000	1,937,899
Silicon Graphics	Mountain View	1,481,600,000	140,700,000	5, 676,591
Raychem	Menlo Park	1,461,500,000	1,700,000	not listed
Solectron	Milpitas	1,456,800,000	55,500,000	2,651,567
Cisco Systems	San Jose	1,243,000,000	314,900,000	15,028,375
Maxtor	San Jose	1,152,200,000	(257,600,000)	not listed

Sources: Data compiled from "Silicon Valley 150," 1995, p. 8G; "What the Boss Makes," 1995, p. 2E. The latter is a list of the top 100 executives in Silicon Valley, measured by total compensations, including stock options, in 1994. The *Mercury News* found no executives earning more than $1,222,791 at the seven firms marked with "not listed."

years ago, some of the Valley's most influential leaders, including the Joint Venture Silicon Valley organization, warned of the Valley's demise. In 1996, they were considerably more upbeat.

Since the invention of the first integrated circuits about 1960, designers and production engineers have squeezed more and more circuit elements (transistors) onto a single chip while reducing the cost of each element. In fact, every two or three years, the number of transistors squeezed onto each chip has quadrupled. Moore's law, named after Intel founder Gordon Moore, suggests that the average cost per circuit element falls about 30% each year. Inevitably, the process will meet up with the physical limitation of Heisenberg's Uncertainty Principle, but that is not on the horizon.

By the early 1970s, with the invention of both the memory chip and the microprocessor, the chip was no longer just another component. It embodied the key functions of computers and other electronic devices. With each new generation of chips, a new array of products or applications emerged, spawning new companies, while often (admittedly not very often) old ones fell by the wayside. The dynamic nature of the U.S. electronics industry as a whole, and the emergence of Silicon Valley as the leading technology center for computers and telecommunications equipment, not just semiconductors, is the result of the growing sophistication of chips.

In 1979, IBM reluctantly joined the emerging trend toward desktop machines and launched its own PC, the most successful family of computers in history. Although IBM by no means abandoned its bread-and-butter mainframe products, the way in which it developed the PC marked a turning point in the structure of high-tech industry. IBM, long a vertically integrated corporate giant, had developed and produced its own chips, keyboards, disk drives, other components, and system software. It had maintained a close relationship with its customers, providing maintenance services and application software.

When IBM introduced the PC, however, it went outside for key elements, obtaining the system software from a small, young Seattle firm called Microsoft and its microprocessor from Intel. Thus, the IBM PC circuit board, with its Asian-assembled chips, looked like what "Big Blue" had become—an international division of labor.

Production employment in Valley high tech has been suffering a long-term decline relative to overall activity, but it is not yet clear if the recent absolute decline results primarily from short-term cyclical factors. It is possible, despite emigration, automation, and a shift to value-added

367	Electronic Components & Accessories	150
381	Engineering and Scientific Instruments	14
382	Measuring and Controlling Devices	21

investment in software, that production employment will again rise with growth in the national economy. Semiconductor wafer fabrication jobs will be held down, as will mass production of many high-tech products.

A SHIFT IN FOCUS

What one sees in Silicon Valley, however, is a shift, a gradual move away from the production of mature products and the fabrication of semiconductors. As companies get larger, they develop the capability to relocate various elements of production and operations to different parts of the country or, for that matter, the world, where production can be done more efficiently or more cheaply. Since almost the beginning of the semiconductor industry, bonding and labor-intensive assembly has been done in Hong Kong, the Philippines and, until recently, Malaysia. Now, wafer fabrication has also moved from Silicon Valley to places like Austin, TX, Albuquerque, NM, and parts of Oregon. As they grow in Silicon Valley, computer companies similarly tend to relocate production to sites where they can pay lower rents, wages, and other reduced production costs.

Silicon Valley is gradually becoming a headquarters and laboratory area, with a reduction of the relative size and importance of the blue-collar workforce. The Valley is still the best location in the world for investment in startup companies and for foreign companies, including those from Japan, South Korea, or Taiwan, wishing to take advantage of available technology. It also remains an excellent place for locally grown companies to introduce new products. When companies introduce new products they want to sell the first versions, the prototypes, locally. What has emerged is a Valley where white collar work has expanded rapidly but where the blue collar component still has a presence.

The difference between the semiconductor industry and computer companies is that when the former dominated the Valley's production, they had a need to closely manage every aspect of the operation through a sequence of extremely fine steps. With submicron architectures, the semiconductor industry must control contamination and power supplies, factors much easier to deal with in house.

The types of production jobs—those dealing with specialized equipment and new products—that remain in Silicon Valley will continue to play an important role in the economy. Whether performed in-house or at subcontractors, these jobs are not likely to disappear. In essence, what could be transferred elsewhere is already gone.

For the past three decades, unions have done relatively little to organize nationally the production/service workers in Silicon Valley. In recent years, the U.S. union movement as a whole has been inclined to believe that Silicon Valley essentially has solved its labor problems and now serves as a vision for the future. This view holds that electronics workers make good money and relax at company after-work athletic facilities in an era of enlightened "postindustrial" America.

Such an idealized vision is not shared by all. For example, the Service Employees International Union (SEIU), which has been organizing janitors contracted by Apple Computer, Hewlett-Packard, and Oracle, sees many labor problems but has shown that strategies exist capable of organizing the most isolated workers in the electronics industry.

A DYNAMIC VALLEY

Silicon Valley, representing as it does the greatest concentration of high-tech talent in the world, is a vital, dynamic, expanding industrial center that is not only profitable but is an increasingly influential location in chip design (semiconductors). The design of the chip, in turn, is becoming the determining design of the computer and chip software the core value-added element in the electronics industry as a whole. The dynamism of Silicon Valley is based in part on the fact that it is still physically possible to squeeze more circuit elements or transistors onto a flake of silicon.

Although electronics has grown in the Valley, employment in the semiconductor industry has essentially leveled off as the industry has learned to do more with less silicon. What the industry has created with each new generation of chips is the possibility for new applications, which, in turn, mean new companies building them or old companies, if they are fortunate, adapting to the new technologies and producing the hardware and software as well. However, it has created a situation in which an important segment of the economy grows to some extent by creating services that were never there before and to some extent because it is providing a way to automate work being done by people elsewhere. To be sure, Silicon Valley is not falling apart. It is doing better on average than the national economy. Some companies are having some problems, but high tech and the Valley itself are very viable sectors of the economy.

FARMING OUT THE "DIRTY WORK"

Currently, however, with major production in the Valley done, for example, by Sun Microsystems, it is very easy to contract out the fabrication of circuit boards, which are standardized. Because workers can be easily recruited and paid a lot less for circuit board assembly, companies such as Sun Microsystems prefer to subcontract the work to Solectron, SCI, or a whole range of other companies that actually produce the computers marketed under Sun's name. Sun, itself, has very few manufacturing personnel. Critics call this "farming out the dirty work."

This kind of work not only exploits cheap labor, but it is also the industry's biggest source of environmental pollution and workplace health hazards. If responsibility can be shifted to subcontractors, major corporations can avoid the onus of social responsibility, associated financial costs, and other consequences. If production somehow lags as a result, the corporation can always pass on the costs to the subcontractors.

Benefits packages present other problems for the industry. In competitive markets, employers wishing to attract the best software engineers have to provide A-1 health care plans, sparking vendors of such plans to insist that all employees be covered. It is at that point that employers farm out work to contractors or subcontractors to avoid providing equal benefits for the entire workforce.

The use of contractors or subcontractors, including temporary and contingent workers without benefits, represents another way in which Silicon Valley is pioneering U.S. industry, actually regressing to a time when workers had almost no control over the relationships and means of production.

WORKING IN THE STRATIFIED VALLEY

Subcontracting has made worse the already long history of stratification of the Silicon Valley workforce. Silicon Valley companies indeed do pay engineers fairly well, as they do others with skills and talent and who meet age, race, and gender conditions. Yet, those who carry the heaviest load of production or service workers, such as janitors, are not treated so well.

What has emerged in Silicon Valley is an intensely stratified workforce, with white (European ancestry) and Japanese-American men, holders of engineering degrees or managerial posts, at the top, and non-white women with no bargaining power in the economy at the bottom. Although percentages in these lower positions are lower locally than they used to be (on a global scale, the numbers are about the same when female electronics workers in Malaysia and the rest of the Third World are added to those in Texas), corporations are hiring essentially the same type of workforce. The difference is that much of the production has been relocated to subcontractors or out of the geographical area entirely.

Extreme discrepancies in income based on race and gender emerge from 1990 census and Equal Employment Opportunity Commission (EEOC) data, the latter of which showed a pattern of white and Japanese-American men at the top and non-white women at the bottom of the ladder. A large percentage of white women is concentrated in clerical work, situated between factory and professional categories. Asians, overrepresented in high-tech industries, are underrepresented in sales, which matches the pattern of stereotypes the industry has about Asians ("High-tech employment," 1992; "Silicon Valley's workforce," 1990).

Combining official EEOC census data with my earlier studies, cited earlier, that used the same dataset reveals an enormous increase of Asians throughout the industry, in both professional and production positions. To a large degree this reflects the hiring preferences of employers, who perceive Mexicans, for example, as more likely to be organized and to give employers problems, whereas Asians tend to be regarded as much more docile.

Not only do few blacks live in Silicon Valley, but the industry has chosen not to spread into areas historically black, even though land values are attractive. High-tech firms have started to move behind the East Bay Hills but not beyond Fremont, or toward Oakland, perceived as a largely black area. They have also stayed out of East Palo Alto, a historically black residential area near the heart of Silicon Valley. The only expressed interest in high tech locating in the area has occurred with what was once rather extreme "sweep everybody out" urban renewal (removal) proposals. The industry has a widespread bias against blacks as "trouble-making" workers, part of a set of stereotypical cultural beliefs that white employers discriminately use in hiring practices.

A third set of information from the 1990 census data is also very illustrative. To work in the industry is to understand that "Asians" is, in fact, a very complex category of peoples, enormously diverse from culture to culture, and that the patterns of Asian migration to Silicon Valley stem from varying conditions and motivations. Disaggregating the census data on Asians reveals, for example, that Filipino-ancestry men rank alongside all women in terms of average income and Japanese-ancestry men actually make more on average than white men. In general, the workers at the very lowest income levels are Southeast

Asians and Mexicans and the small percentage of blacks, while those in top wage categories are Northeast Asians and Indians with backgrounds in engineering. There are relatively few South Asians in the Valley, but those involved as software engineers tend to be high income earners and near the top of their regional grouping. The concept of Asian, previously understood to be average income earners, therefore breaks down into far more complexity and differentiation with more national specificity.

Understanding these demographics and the business culture in Silicon Valley suggests a long-held belief that the industry is less about New Age meritocracy and more about race and gender discrimination. When these data are published (see, e.g., *Global Electronics*), the response of the industry and its apologists is something along the lines of, "Well, of course, there are not that many women engineers, and there are not that many women who are trained to run high-tech companies." In fact, a number of studies do show that a major explanation for the lower income of women is that they lack training, but another big factor is discrimination. Systematic discrimination and unequal education are realities that must be faced, just as the fact that the relatively few Chicanos with training in computer science cannot ignore the cultural, financial, and political obstacles that keep them out. When a number of studies have shown that equally trained and experienced women still do not get paid as much as men, industry has to respond to the fact of discrimination (see, e.g., Fernandez, 1993; Levander, 1991).

Looking beyond high-tech industrial employment to the janitors, the food-service workers, the landscapers—people who are part of the high-tech economy in Silicon Valley but not actually doing high-tech work—one finds a preponderance of Hispanic workers. The overrepresentation of Asians in assembly work does not spill over into these other services. Most janitors are Hispanic, and although they represent a large number of people in the region, clearly they do not reap the benefits of the Valley's economic success.

ETHNIC AND GENDER STRATIFICATION

By definition, high-tech industry employs a disproportionately large number of professionals, but Silicon Valley, where most labs and headquarters are located, hosts an even greater share of engineers and computer scientists. The low end of the workforce is smaller in Silicon Valley than elsewhere, but the problems of working in electronics are serious, both for the tens of thousands of local workers and the hundreds of thousands worldwide. The high-technology industries have brought a sharp polarization of income and security in Silicon Valley, and as its products spread throughout the economy, it is sharpening those differences elsewhere.

Many factors contribute to the race and gender segregation of the high-tech workforce (see figure 5.1). Education is an important factor, because most of the high-level employees have one or more college degrees, and such technical credentials tend to favor white and north Asian males. Among wage workers, companies often practice more overt race and gender discrimination, and although not openly admitting to such in the United States, where it is illegal, they routinely discriminate when hiring at their overseas plants. In their Asian production sites, employers specifically advertise for young, unmarried women, who generally are more easily exploited than adult males (see Table 5.2).

In the Valley's electronics manufacturing industry, white men make up 62.8% of the officials and managers and an even greater portion of corporate officers - 50.9% of all professionals. With the exception of a growing number of underemployed older engineers, these technical specialists earn good money and are in a position to enjoy it. With an average family income approaching $54,000, Santa

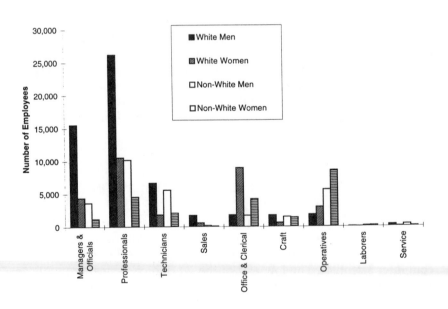

Figure 5.1. Job type by race and gender, 1990

	Population	Number of Employees										% Pop.	Percentage of Job Categories									
		Total	Mgrs	Profs	Techs	Sales	Clerks	Craft	Oper	Lab	Serv		Total	Mgrs	Profs	Techs	Sale	Clerks	Craft	Ope	Lab	Serv
ALL	1,497,577	136,909	24,737	51,468	16,078	2,580	16,598	5,278	18,951	374	845	100.	100.	100.	100.	100.0	100.	100.	100.	100	100	100.
Men	759,503	84,699	19,158	36,373	12,191	1,911	3,474	3,264	7,452	188	688	50.7	61.9	77.4	70.7	75.8	74.1	20.9	61.8	39.	50.	81.4
Women	738,074	52,210	5,579	15,095	3,887	669	13,124	2,014	11,499	186	157	49.3	38.1	22.6	29.3	24.2	25.9	79.1	38.2	60.	49.	18.6
Whites	869,874	85,739	19,902	36,747	8,460	2,297	10,688	2,377	4,817	69	382	58.1	62.6	80.5	71.4	52.6	89.0	64.4	45.0	25.	18.	45.2
Men		55,820	15,525	26,198	6,638	1,722	1,773	1,754	1,841	50	319		40.8	62.8	50.9	41.3	66.7	10.7	33.2	9.7	13.	37.8
Women		29,919	4,377	10,549	1,822	575	8,915	623	2,976	19	63		21.9	17.7	20.5	11.3	22.3	53.7	11.8	15.	5.1	7.5
Non-		51,170	4,835	14,721	7,618	283	5,910	2,901	14,134	305	463		37.4	19.5	28.6	47.4	11.0	35.6	55.0	74.	81.	54.8
Men		28,879	3,633	10,175	5,553	189	1,701	1,510	5,611	138	369		21.1	14.7	19.8	34.5	7.3	10.2	28.6	29.	36.	43.7
Women		22,291	1,202	4,546	2,065	94	4,209	1,391	8,523	167	94		16.3	4.9	8.8	12.8	3.6	25.4	26.4	45.	44.	11.1
African-	52,583	5,495	600	1,429	767	49	1,221	296	1,051	16	66	3.5	4.0	2.4	2.8	4.8	1.9	7.4	5.6	5.5	4.3	7.8
Men		2,938	407	910	583	35	360	179	404	9	51		2.1	1.6	1.8	3.6	1.4	2.2	3.4	2.1	2.4	6.0
Women		2,557	193	519	184	14	861	117	647	7	15		1.9	0.8	1.0	1.1	0.5	5.2	2.2	3.4	1.9	1.8
Hispanic	314,564	12,753	1,062	2,071	1,811	77	2,254	973	4,096	134	275	21.0	9.3	4.3	4.0	11.3	3.0	13.6	18.4	21.	35.	32.5
Men		6,201	710	1,306	1,266	51	581	584	1,422	61	220		4.5	2.9	2.5	7.9	2.0	3.5	11.1	7.5	16.	26.0
Women		6,552	352	765	545	26	1,673	389	2,674	73	55		4.8	1.4	1.5	3.4	1.0	10.1	7.4	14.	19.	6.5
Asians	251,496	32,349	3,084	11,066	4,947	149	2,329	1,599	8,901	154	120	16.8	23.6	12.5	21.5	30.8	5.8	14.0	30.3	47.	41.	14.2
Men		19,425	2,458	7,848	3,638	98	740	731	3,749	67	96		14.2	9.9	15.2	22.6	3.8	4.5	13.8	19.	17.	11.4
Women		12,924	626	3,218	1,309	51	1,589	868	5,512	87	24		9.4	2.5	6.3	8.1	2.0	9.6	16.4	27.	23.	2.8
Native	6,694	573	89	155	93	8	106	33	86	1	2	0.4	0.4	0.4	0.3	0.6	0.3	0.6	0.6	0.5	0.3	0.2
Men		315	58	111	66	5	20	16	36	1	2		0.2	0.2	0.2	0.4	0.2	0.1	0.3	0.3	0.3	0.2
Women		258	31	44	27	3	86	17	50	0	0		0.2	0.1	0.1	0.2	0.1	0.5	0.3	0.3	0.0	0.0
	2,366											0.2										

This table was prepared by Lenny Siegel of the Pacific Studies Center from data provided by the U.S. Equal Employment Opportunity Commission. Population figures are from the 1990 U.S. Census. The figures include high-tech manufacturing, but not services.

Terms: Silicon Valley = Santa Clara County; Mgr. = Officials & Managers; Prof. = Professionals; Techs = Technicians; Sales = Sales Workers; Clerks = Office & Clerical Workers; Craft = Blue-Collar Skilled Production; Operatives = Semi-Skilled Blue-Collar; Labor = Laborers/Unskilled Blue-Collar; Service = Service Workers

Table 5.2. High-Tech Manufacturing Employment in Silicon Valley, CA-1990

Clara County is the state's wealthiest metropolitan area and ranks seventh in the nation.

White men only account for 9.7% of plant operatives. Women of all races make up 79.1% of the clerical workers and 60.75% of the operatives, yet they make up only 38.1% of the high-tech manufacturing workforce. Non-white women account for 45% of the operatives, although they only represent 16.3% of the workforce at the same companies. Non-whites (men and women) account for 74.6% of the operatives.

Asians (including immigrants and Asian Americans) are highly represented in most Silicon Valley job categories. Although they make up 16.8% of the area's population, they represent 23.6% of the high-tech manufacturing workforce. Their share of technicians (30.8%), craft workers (30.3%), semi-skilled operatives (47%), and unskilled laborers (41.2%) is even greater than their rate of participation in the high-tech workforce, and they also make up a substantial fraction of the professional workforce (21.5%).

High-tech employers like to hire Asian professionals, who are often well educated. Chinese and Japanese of both genders, as well as Korean and Indian men, earn high pay on the average, yet there is a glass ceiling that is likely imposed on the basis of ethnic prejudice. Only 12.5% of the managers and 5.8% of the sales workers are Asian. A recent survey of Asian-American white collar workers in the Valley found that 58% "knew Asian-Americans who believed they were denied a promotion because of race" (Asian-Americans for Community Involvement, 1993).

Hispanics, however, are not well represented in the high-tech industry. They make up 21% of the county population, but according to the Equal Employment Opportunity Commission, they represent only 9.3% of the electronics manufacturing workforce. Although Mexicans and Mexican Americans account for 6.7% of the entire employed Valley workforce, only 4.4% of the electronics workers in Santa Clara County are Mexican. In fact, three distinct Asian ethnic groups—Chinese, Filipinos, and Vietnamese—each outnumber Mexicans, who still represent the largest non-white ethnic group in the overall workforce.

To some degree, the high number of Asians is attributable to their average higher level of technical education, particularly among Northeast Asians and Indians, but also to their being perceived (especially Southeast Asians) as more docile and less inclined to individual militancy or unionization than Mexicans. The belief among managers is that Filipinos and Vietnamese are culturally (or even biologically) more deferential than Mexicans and Central Americans. The militancy seen in some Southeast Asian factory settings such as in the Philippines or Thailand (especially among women workers) belies such a stereotype. What is relevant here is that the United States has virtually no control over who enters the country from Mexico, whereas Asian immigration tends to favor middle-class entrants. Hispanics are considered good maintenance workers. At electronics firms reporting to the EEOC, 32.5% of the in-house "service" employees are Hispanics, and most contract janitors who clean up high- tech firms are Mexican. By organizing with the Service Employees International Union, these workers have reinforced managers' notions of Mexican militancy.

Differences in status show up in annual pay levels, as reported in the 1990 Census Public Use Microdata Samples (PUMS). Analyzing PUMS data covering Santa Clara county, the southern reaches of neighboring Alameda and San Mateo counties, and workers in other area counties who commute to Santa Clara County, the Pacific Studies Center found that in virtually every major ethnic group, men earn much more than women[2] (see Figure 5.2).

Pay differentials among racial groups are also large, and using census data, contrasts among Asian ethnic groups—hidden by the combining of Asian groups in other labor statistics series—are apparent. Japanese-Americans, as well as Chinese Americans and immigrants from Taiwan and China, earn incomes on par with whites, whereas Southeast Asians, including Filipinos and Indochinese, have earnings about the same as Mexican and other Hispanic groups.

In this study, the Pacific Studies Center averaged the annual earned income of individual workers, not family or household. Differences, therefore, could result from: (a) different pay rates (wage or salary), (b) varying lengths of workweek (hours), and (c) duration of employment during the year. The analysis did not determine whether pay differentials are due to varying educational backgrounds, differing

---

[2]The PUMS data include information from 5% of the total census sample. The employment counts shown in Table 5.3, therefore, are the actual sampling multiplied by 20. The ethnic data are based on the "nation of origin" questions. We have been somewhat arbitrary in applying the suffix "American," thus, Mexican Americans could either be Mexican immigrants or native-born Americans of Mexican descent. Income is based on the arithmetic mean of earned income of individual workers, as reported voluntarily by the individuals. It is supposed to be an actual annual figure for 1989. Note that the other employment data, such as the industry in which the respondent worked, were valid as of the date of enumeration in 1990. "High-tech," as used in the census study, includes the following census categories: Computers & Machinery, Equipment, and Supplies, not elsewhere classified; Not Specified Electrical Machinery, Equipment and Supplies; Scientific and Controlling Instruments; Guided Missiles, Space Vehicles, and Parts; and Computer and Data Processing Services.

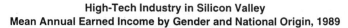

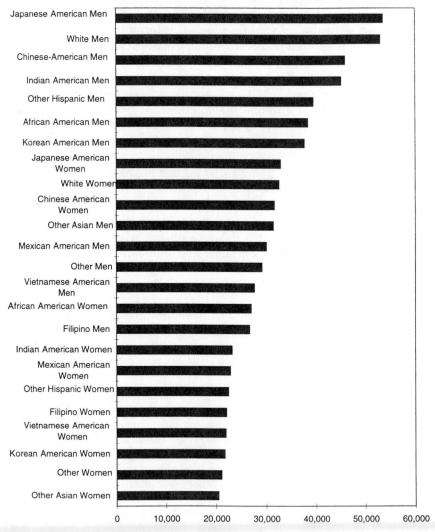

Figure 5.2. High-tech industry in Silicon Valley (mean annual earned income by gender and national origin, 1989)

	Electronics		High-Tech		High-Tech/Related Services		All Industries	
	Number	Mean Annual Income	Number	Mean Annual Income	Number	Mean Annual Income	Number	Mean Annual Income
White Men	75.420	$54.03	102.460	$52.999	109.840	$51.65	409.900	$41.663
White Women	42.740	$32.24	53.540	$32.594	63.380	$30.25	332.180	$22.724
Mexican-American Men	4.440	$28.04	5.240	$30.037	7.420	$26.13	46.980	$20.641
Mexican-American	4.740	$22.14	5.300	$22.821	6.840	$19.80	30.580	$15.122
Other Hispanic Men	1.720	$39.19	2.000	$39.474	2.340	$35.57	12.000	$28.385
Other Hispanic Women	1.760	$21.21	2.000	$22.430	2.340	$22.40	11.620	$18.315
African-American Men	3.940	$35.63	4.800	$38.399	5.540	$35.52	21.960	$27.569
African-American	3.060	$25.97	3.540	$26.983	4.180	$25.08	18.740	$20.706
Chinese-American Men	10.360	$45.60	12.360	$45.891	12.660	$45.35	28.060	$38.111
Chinese-American	5.460	$30.67	6.500	$31.699	6.840	$30.76	20.800	$24.676
Filipino Men	7.680	$25.40	8.760	$26.671	9.180	$26.20	26.280	$23.894
Filipino Women	8.220	$21.88	8.700	$22.028	9.200	$21.52	27.140	$20.551
Japanese-American Men	2.700	$54.72	3.280	$53.531	3.500	$51.39	10.220	$43.329
Japanese-American	1.440	$33.64	1.860	$32.922	2.100	$30.19	9.820	$23.494
Indian-American Men	4.500	$43.67	5.040	$45.090	5.160	$44.64	10.200	$37.515
Indian-American Women	1.640	$23.01	1.840	$23.147	1.900	$22.85	5.500	$18.507
Korean-American Men	1.380	$38.40	1.500	$37.695	1.660	$36.68	4.100	$30.761
Korean-American	1.360	$21.61	1.380	$21.696	1.460	$20.72	4.260	$18.792
Vietnamese-American	6.860	$27.21	7.220	$27.630	7.440	$27.24	15.320	$23.872
Vietnamese-American	4.280	$21.89	4.480	$21.890	4.640	$21.51	9.980	$17.848
Other Asian Men	1.380	$31.45	1.380	$31.451	1.420	$30.97	3.480	$27.050
Other Asian Women	1.020	$20.42	1.040	$20.431	1.040	$20.43	2.540	$17.943
Other Men	6.280	$28.08	7.240	$29.149	9.920	$25.68	56.100	$21.632
Other Women	5.360	$20.55	6.140	$21.044	7.780	$19.26	40.400	$15.947
TOTAL	206.640	$39.43	256.500	$40.283	286.680	$38.17	1,157.06	$29.610

Source: U.S. Census Public Use Microdata Samples

Table 5.3. Silicon Valley Employment by National Origin and Gender, 1990 Silicon Valley Annual Earned Income by National Origin and Gender, 1989 Source: Data compiled from U.S. Census Public Use Microdata Samples.

skill levels, discriminatory employment practices, or other factors. The census did not ask whether employees chose part-time work or voluntarily worked only a portion of the year, so it was not possible to determine whether any group's low income was by choice.

Regardless of the reasons for the income inequality, variances are large enough to confirm that the earning power of distinct ethnic and gender groups is socially significant. Any policy designed to promote high-tech industry as a solution to the United States' economic woes must take those differences into account, and subsidies to industry must be based on their contribution to society. Labor law should not impede the workers' right to organize into unions, and educational programs, particularly those that promote computer literacy, should target females and traditionally low-income ethnic groups.

TWO VALLEYS

There are essentially two Valleys, two worlds of people who work in the same physical area but live along great financial, cultural, and geographical divides. Although some engineers drive considerable distances, it is mainly the lower paid workers living in communities such as San Jose, Gilroy, or the East Bay who are forced to make long commutes. Engineers are privileged in that they live close to the suburban plants (a pattern established before revelations about the toxic threat of semiconductor production). The cost of housing in the Valley has been driven up by an influx of professionals attempting to avoid increasing commuter costs. People priced out of these housing markets get to do the heavier commutes.

Historically, industry has been concentrated in northern Santa Clara County: Palo Alto, Mountain View, Sunnyvale, Santa Clara, and Cupertino. Those communities play host to the companies during the day and the better paid, high-tech workforce at night. They have a tax base shored up by corporate property and sales taxes. When a company such as Apple sells computers, the communities listed as the point-of-sale benefit from enormous sales taxes. In fact, the tax base in the northern Valley cities is proportionately much bigger than in such southern areas as San Jose, which serves the lower income residents.

Mechanisms are in place that intend to balance the impact on education from the differentials in the tax base, but generally they do not work completely. The income differentials appearing in the census data are, in part, geographical in nature. In poorer, outlying areas, industry residents are not getting the value-added benefits created by the companies in which they work, for the historic pattern has been of economic concentration near the center of the Valley. The problem, however, is not quite that simple. There are new firms building up in Fremont; there are professionals commuting from Manteca. However, the urban pattern in this part of California is different from traditional ones elsewhere in the United States, where white-collar workers commuted the greatest distances and lived in wealthy suburbs. In the Valley, the factories are located in the wealthy suburbs, and it is the big city, San Jose, that suffers from underinvestment.

ORGANIZING THE VALLEY

What is to be done? First, one has to try to influence urban planning. I have been arguing that with the Navy pulling out of Moffett Field (in the Bay Area), it should be converted to provide housing for people who work in the nearby electronics industry. What are needed are rezoning regulations that help make higher density, less sprawl-type, housing availability, so that production workers can actually live close to work. That is one important way to address the problem, but the key element is income distribution. "And to many," according to one observer, "[SEIU Local President Mike] Garcia's [labor organizing] crusade is an effective, dignified anti-poverty program" (Rafferty, 1993, p. 10). Whether they work directly for Apple, Hewlett-Packard, Sun, Solectron, Shine (the janitorial contractor for Apple), or any of the subcontractors, workers at the lowest levels need to be organized to benefit from the wealth generated by high tech investment in Silicon Valley.

This will not be easy. Organizing in the Valley has to be industrywide because worker mobility is a safety valve for the companies: When workers are unhappy, they can switch jobs. This is true at all levels. Organizing also has to be backed by significant resources to deal with problems such as that of Versatronex, where workers organized and the company decided to shut down. Workers are caught in a competitive marketplace and are having to sell sweatshop labor at sweatshop prices.

The approach that the janitors' union has taken, using all types of pressure against brand-name businesses (Apple, Hewlett-Packard, etc.) and holding them accountable for the labor practices of subcontractors that operate cleaning services on their property, seems to work. It is successful, in part, because the distraction and cost of bad public relations is a bigger burden than small increases in pay drawn from substantial revenues pools for the relatively few janitors. This is an approach that other unions, including those in manufacturing and food service, need to consider.

At Sun, Hewlett-Packard, and Apple, unions need to say, "You claim you treat your workers well, but we want you to take responsibility for the conditions of the workers of your subcontractors. You are farming out the 'dirty work.'" The subcontractors cannot simply shut down because, under pressure, the main company—the one that is visible in the media, that is susceptible to consumer boycotts, and the one whose corporate image matters—will have to pay the price of solving those kinds of problems. That is what happened at Apple with the janitors' organizing campaign. When Hewlett-Packard saw what happened at Apple, they quickly came to terms with the janitors.

THE ORACLE EXPERIENCE

Oracle has long been under pressure to collect union representation cards and, if necessary, to recognize the union. Oracle claims, however, that because it does not own its buildings, it has nothing legally to do with the people who clean the offices. In ethical terms, it is hard to justify the claim that the people who come in and clean the offices as soon as the white collar employees leave have no relationship to that company.

The janitors had tried to get union recognition by Service Medallion, the cleaning service firm that cleans Oracle's buildings. With the help of a small number of people working inside Oracle, who were somewhat sympathetic to the janitors, union activists located e-mail addresses of key Oracle employees and started sending them messages:

> You can support the janitors' struggle just by asking the Oracle Corporation if they really intend to support exploitation under their own roof. Although Oracle has so far refused to consider the workers' requests for fair treatment, change is possible. Programmers, engineers, and other skilled high tech workers can play a critical role by bursting the bubble of invisibility that allows the abuse to go unchallenged.

Friends of union organizers using e-mail from campuses around the country also were encouraged to send Oracle's top management messages of support for the workers. Management reacted by sending their own e-mail messages to all company employees. They had no idea how many people the union activists had reached in the company, nor, apparently, did they monitor the situation.

Management responded with "We are not beating our janitors"-type messages. Their defensive tone got people to start asking questions, which drew attention to and created a climate of heightened awareness about the janitors' situation and led to demonstrations, although it did not capture media attention. There were not hundreds of people picketing outside the gates. However, the cost of negative publicity and lowered morale, induced by the e-mail strategy, was significant, whereas the cost of paying $1 more per hour to fewer than 50 people was not. There is no way of knowing exactly what impact e-mail had in this case, but to activists who use that communication process, the prospects of penetrating political barriers are exciting. In the long term, companies will set up filters to block unwanted messages, but there will always be new communication measures and countermeasures.

The key is to use the organizing tactics of the janitors at Oracle in other areas of high-tech work: Go after the "name" companies, put pressure on them, collaborate with various community ethnic groups, and bring in Asian organizers, for example, Vietnamese-speaking organizers to work with the Vietnamese. This approach, which has a good chance of success, will require a lot of resources, but, looking at the workforce profile, there is no choice.

THE NEW MIDLEVEL JOB ELEMENT

If one studies the testing and customer support needs of the computer industry, it is clear that new midlevel employment has to increase. Whenever new hardware or software applications are introduced, they must be system-compatible and tested as such, which is an increasingly complex task. With more people using computers with different applications, there is more need for customer service, a growing and new midlevel employment area. The industry tends to be top and bottom heavy in employment, with very few skilled blue collar workers such as one finds in auto plants or aircraft factories, in which blue-collar workers are paid $15 to $20 an hour.

Because midlevel software work is new, it has not become a national phenomenon and is likely to occur only in areas such as Silicon Valley or Seattle, with their concentrations of software businesses. It is important in educational planning that the needed preparation be available to females and children of color. If in the schools there is not broad access to computers, and opportunities for computer literacy are largely restricted to middle- and upper-class white males, as has been the case historically, the new midlevel jobs will not go to women and ethnic minorities in any significant numbers. Children of Japanese and Chinese ancestry stand the best chance to compete with whites in this job sector.

One question is, however, to what degree are these midlevel positions being "proletarianized"? Are people in customer service/support and software quality assurance finding that in their work they no longer have, like engineers, a seller's market? Is their pay being driven down, and are they being forced to work as temporaries, with less job security? If so, can they also be organized? Midlevel software workers might not prefer unions, but they might conceivably want a guild or other type of organization to help address some of their employment needs that associations traditionally have done for engineers in other industries.

As organizing expands into high-tech localities, new tactics will be used. Creative uses of the very technology of the industry can become a powerful tool to undermine the social relationships that have been built under it. That is the dialectical challenge that union activists must learn.

REFERENCES

Asian-Americans for Community Involvement. (1993, September 10). Qualified but. . . . A Report on glass ceiling issues facing Asian Americans in Silicon Valley. *San Jose Mercury News,* p. 1A.

Datamation 100. (1995, June 1). *Datamation,* p. 69-89.

Fernandez, M. (1993). *The gilded cage of professional life: Santa Clara Valley's people of color face the glass ceiling.* (Unpublished paper). Santa Clara, CA: Santa Clara University.

The Fortune 500 ranked within states. (1995, May 15). *Fortune,* p. F32.

High-tech employment patterns in Silicon Valley, 1990. (1992, October). *Global Electronics.*

Levander, M. (1991, June 9). Survey: Businesswomen feel left behind. *San Jose Mercury News,* p.10.

Rafferty, C. (1993, September 12). Its a dirty business. *West Magazine, San Jose Mercury News,* p. 10.

Silicon Valley 150. (1994, April 10). *San Jose Mercury News,* p. 16.

Silicon Valley's workforce remains segregated. (1990, February). *Global Electronics.*

What the boss makes: The Valley's top 100. (1995, June 19). *San Jose Mercury News,* p. 2E.

6

Electronics, Communications, and Labor: The Malaysia Connection*

Gerald Sussman

Who built the seven towers of Thebes?
The books are filled with the names of kings.
Was it kings who hauled the craggy blocks of stone?...
In the evening when the Chinese wall was finished,
Where did the masons go?
    —Bertolt Brecht (quoted in Brier, 1992, p. ix)

Although people like Steve Jobs, David Packard, and Edward Roberts often get the credit for developing the contemporary computer industry, the industrial workforce comprising mostly women, in places like Penang's "free industrial zone," Mexico's maquiladora factories, Sony electronics in Wales, and the Bataan Export Processing Zone in the

*I express my appreciation to the Association for Asian Studies for providing a Luce Foundation grant that enabled me to conduct research in Malaysia in August and September 1995. I also thank my hosts in Malaysia at the Universiti Sains Malaysia in Penang and the Mara Institute of Technology in Shah Alam, especially Ramli Mohamed, Hamidah Isri, and Sankaran Ramanathan, and the

Philippines, are the real "masons" of the "digital revolution." The big electronics firms in the United States and Japan look primarily to the Third World to supply components needs, while the ability of computer corporations such as IBM and Apple to provide digital office equipment and desktop PCs relies on the micro component assembly in Malaysia of companies such as Intel, Motorola, Hewlett-Packard, Seagate, and others.

Half of Apple's desktop computers and the largest share of disk drives are made in Singapore, and 60% to 70% of its components are made in various parts of East and Southeast Asia. Taiwan produces 80% of the world's computer motherboards and a large percentage of their monitors. Memory chips are manufactured in South Korea. Keyboards are produced in Thailand. The leading computer component manufacturer, Seagate, makes 80% of its parts in Malaysia, a country that is also a leading producer of floppy drives. According to one observer, "pry open an American brand desktop personal computer at a Radio Shack outlet in, say, New York, and the only two items marked 'made-in-U.S.A.' may be the microprocessor and the label. Every other part probably comes from somewhere in Asia" (Thornton, 1995).

This chapter discusses Malaysia's role in the production of communications and information technologies, focusing on the contributions and conditions of its workforce in computer, industrial, and consumer electronics. The purpose is to show the kinds of social subsidies that go into the production of electronics for products used in the core industrial countries. With its extensive poverty and low labor costs (one tenth of those of Japan and the United States), an English-speaking workforce (owing to its colonization under Britain), relative political stability, and good infrastructure, Malaysia was targeted in the late 1960s as a desirable location for electronics investment. In Penang's Bayan Lepas free trade zone (FTZ; later renamed free industrial zone following 1990 amendments to the 1971 Free Trade Zone Act), rural workers were induced to migrate from the neighboring states of Perlis, Kedah, and Perak in hopes of alleviating the poverty of their parents and siblings, working three shifts and at boom periods for 12 hours at a stretch (Tan, 1986).

The rehabilitation of the war-torn western European and Japanese economies by the 1960s created a more competitive world industrial economy, inducing the leading industrial power—the United

many people I interviewed who gave generously of their time and knowledge. Rajah Rasiah, who shared with me much of his expertise and resources, is widely regarded as the most authoritative scholar on the Malaysian electronics industry. John Lent, with his keen editor's eye, gave valuable suggestions. Meiru Liu provided assistance in collecting research materials for this project. The shortcomings of this chapter are entirely my own.

States—to seek out lower cost production sites in the Third World for the manufacture of lightweight transportable commodities, especially in the high tech semiconductor and electronics industries. The postwar development of communication satellites, more extensive international facilities for jet aircraft, and new containerized shipping technology served the infrastructural requirements of dispersed industrial production. Although U.S. capital had an advantageous global position at the end of the war, postwar Japan (and, to an increasing extent, the so-called Asian newly industrializing countries, ANICs) required more direct state sponsorship and foreign assistance to transform their respective and limited manufacturing bases into higher-end, technology-intensive industries. This brought relative improvements in working conditions in Japan, but the ANICs (South Korea, Taiwan, Singapore, Hong Kong) have remained far behind Japan and the West in primary and secondary wages (benefits, etc.).

In Malaysia, the postwar period brought escalating racial tensions between Malays and the generally more affluent Chinese, who dominated domestic capital, commercial, professional, and technical sectors, relations of power that had been structured under earlier British colonial rule. The dominant political alliance—the United Malays National Organisation (UMNO)—together with its weaker partners— the Malaysian Chinese Association and the Malaysian Indian Congress—was able to exploit racial tensions in order to solidify its base of power under a rubric of communalism that required the government to address income disparities. With the eruption of racial tensions in 1969, UMNO's communalist-oriented "special rights" policies for Malays (providing "affirmative action" quotas in employment, imposing language and marriage restrictions, and opening privileged opportunities in civil service, scholarships, grants, licenses, credit, and loans) did, in fact, strengthen the state's power base. This was achieved through a class alliance among the Malay, Chinese, and Indian bourgeoisie, who collaborated within the Barisan Nasional (National Front) political formation for mutual gains under the state's newly formed export-oriented "New Economic Policy" (NEP).

As elsewhere in Southeast Asia, Malaysia submitted to the World Bank's program of "export-oriented industrialization" (EOI), organized under the Bank presidency (1968-1981) of former U.S. Defense Secretary Robert McNamara, who sought political economic alternatives both to socialism and to his failed war policies in Vietnam. McNamara developed a Pacific Rim strategy in this direction (George & Sabelli, 1994). EOI offered foreign investment to absorb Malaysia's surplus labor in exchange for attractive state inducements to subsidiaries of transnational corporate (TNC) manufacturers. Of the export-processing zones (also called free trade or free industrial zones) created in that country, Penang's (an island in the northwest of peninsular Malaysia) was the first and remains the largest, and electronics has been Malaysia's and Penang's biggest sectoral investment area. Investing firms were given tax exemptions; subsidized infrastructural supports for land, water, and electricity; easy access to a nearby international airport; and protections for long-term investments, including a ban on unionization in "pioneering industries" (Rasiah, 1994a).

Unlike the ANICs, Malaysia has not lost (1996) favorable Generalized System of Preferences (GSP) tariff status from the United States, the largest market for its duty-free microelectronic exports, despite efforts by the AFL-CIO (U.S.) trade federation to have this status removed. Taiwan and Singapore capital have responded to the loss of their own GSP status by moving some of their plants to Malaysia, joining OECD investments already there (Rasiah, 1989). As part of a growing regional division of labor within East/Southeast Asia, Taiwan and Singapore capital have taken advantage of the relative low wages and abundant labor pool in Malaysia to use that country as a reserve workforce.

Since the late 1960s transnational capital has reorganized the world economy toward a more globally integrated task-segmented system of production, what is now commonly referred to as the new international division of labor (NIDL). Evolved under the pressures of oligopolistic competition and falling rates of profit in the leading capitalist industrial states, the NIDL has created the means for TNCs to increasingly relocate "offshore" and, in consortium with transnational lending institutions, to pressure Third World states to open themselves to private industrial markets, export-oriented industrialization, and export-processing zones for foreign subsidiaries. Of some two million workers in these EPZs by the mid-1980s, two thirds were employed by TNC subsidiaries; a total of 7 million Third World workers were employed by TNCs overall (Kreye, Heinrich, & Frobel, 1988). EOI has also transformed the sexual composition of labor by bringing millions of young women into the industrial laborforce, whose participation in manufacturing rose from 28.1% in 1970 to 46.4% in 1990 (Rasiah, 1955c).

Part of the explanation for NIDL which sets this form of transnational capital transfer off from earlier world economic structures, can be found in the increased technology-based opportunities for accumulation on a global scale, augmented by the commercial spinoffs of huge military R & D spending in transportation and communications (e.g., satellites, launch technology, space vehicles, ground stations, jet aircraft, fiber optic cable, advanced electronics, computer hardware and software, digital switching and transmission systems, high-resolution Earth reconnaissance photography, etc.). These technologies have been put at the disposal of corporate capital to offset cyclical economic downturns and have greatly improved the productive potentials of invested capital, while reducing dependence on location-specific skilled and unskilled labor pools. The 1980s' deregulation of freight hauling, airlines, telephony, and computer industries created wider cost-competitive and spatial control options, allowing AT&T and other telephone corporations, IBM and the computer manufacturing industry, and other communications (and transportation) firms to branch out to new regional and technical markets. Deregulation and a political environment supportive of footloose capitalism in the 1980s also encouraged corporations to relocate capital, labor, and consumer markets overseas, especially to Third World locales.

Flexible, customized, computer-aided design in manufacturing together with telecommunications have enabled TNCs to reorganize production along small batch and "just-in-time" (JIT) output schedules and other cost-reduction management techniques to maximize the fluidity and diffusion of capital over time and space. Innovations of these kinds incorporate strategies to reduce the cost and value of labor and increase market power, productivity, and profit for corporate ownership, while overcoming temporal and locational constraints of the past. For organized labor, NIDL represents a more sophisticated stage of class conflict and a growing threat to Third World political, economic, social, and cultural identity. Malaysian electronics workers, like their Western counterparts in the automobile industry, have had to face a major challenge to the survival of the skilled and unskilled workforce in the face of continuing automation and innovation of the shopfloor.

Since the 1970s, several Third World countries have come into their own as producers of microelectronics parts and devices, integrating their workforces as part of the new international division of labor in the production of various industrial and consumer computers and communication equipment. This is a sector of modern industrial capitalism that the *Financial Times* of London identified more than a decade ago as the "crude oil of the new industrial revolution" (quoted in Chaponniere, 1984, p. 136). Unlike petroleum, electronics production is not location dependent, which means that whereas some countries can expect to remain dominant in the industry well into the future, especially the United States and Japan, other countries that now have lower end manufacturing niches cannot count on remaining "competitive." Although countries such as Singapore, South Korea, Taiwan, and Malaysia are important to NIDL, newer entrants such as China, India, and even Vietnam are poised to sell themselves as the new cheap labor havens for transnational investment in component and software development. By the early 1980s, Southeast Asia was producing 85% to 90% if all offshore production of semiconductor assembly (Scott, 1987).

Along with these specific incentives, the Malaysian government assured the foreign investment community from the outset that it would block "irresponsible" trade unionism, stabilize domestic currency, guarantee a strong foreign currency reserve, follow acceptable fiscal and monetary policies, and assume obligations imposed by the World Bank or the IMF (Hua, 1983). The federal government of Mahathir Mohamad (1981-) has relied on UMNO's "bureaucratic authoritarianism" to protect the interests of the state, its private sector backers, and its relationships with foreign capital, while labor bears the heaviest burden of its developmentalist integration with the world economy. Paid one tenth to one twentieth the rates of their Western counterparts and lacking access to state political processes or the superior mobility of corporate capital (augmented by modern communication and transportation systems), Third World labor as a whole has few independent options other than through militant organizational tactics to challenge ownership and force concessions from the state. Within NIDL-type substage production of communication and information parts, software, and equipment, Third World labor faces fragmentation, marginalization, and large-scale immobility, emerging as the "new helots" of capital (Cohen, 1987).

As a major producer and exporter of semiconductor devices and seeking forward linkages to more advanced production technologies, Malaysia presently occupies a lower middle tier within the new international division of labor in electronics, an aspiring ANIC. One of the specific social characteristics of NIDL, in Malaysia as in other electronics-producing countries, including the leading industrial powers, is that a large majority of assembly workers are female. In Malaysia, this has been a deliberate expedient of the state in collusion with foreign capital, rather than an organic evolution within the country's labor development as a whole. The Malaysian government pitched the idea of the "dexterous hands of the oriental female" to attract TNCs and foreign capital, but the gender politics had a logic beyond considerations of nimble fingers (Salih and Young, 1989). It was understood by the state and foreign capital that the nature of the work would involve highly stressful and dangerous shopfloor conditions, against which women were less likely to organize or rebel.

One of the earliest of the Asian states to adopt EOI (Taiwan was first to set up an "export processing zone" in 1965), Malaysia enacted an Investment Incentives Act for that purpose in 1968, followed by a Free Trade Zone Act in 1971. The 1968 Act created attractive incentives for foreign capital to invest in labor-intensive industrial development, which would concentrate mainly in the country's new EPZs. Electronics and electrical components manufacturers were among the original investors given priority and pioneer investment status. National Semiconductor (U.S.), which had relocated its semiconductor production to Singapore in 1968, was Malaysia's first entrant (1971) under the Act.

The Act led to Malaysia becoming the world's third largest producer (behind the United States and Japan) and largest exporter of semiconductors by the 1980s, with electronics being its principal earner of foreign currency since 1987 (58.2% of manufactured exports by value in 1992). The United States and the European Economic Community (EEC) initially were the major export markets for firms producing microcircuits, transistors, diodes, and conductor devices, but in the 1980s and 1990s, Japan became a leading trade partner in this sector and a leading investor nation. By 1985, component assembly represented 90% of the value added in the industry, and close to 100% of assembly investment was of foreign origin (Fong, 1989; Stockton, 1991). Electronics has been the country's largest single employer, representing 30.7% of manufacturing employment by 1993 and propelling what appears to be a growth trajectory in Malaysia's industrialization process and economic development. In 1970, manufacturing contributed 13.4% to Malaysia's GDP, and by 1990 it was up to 26.6% (Rasiah, 1995b; Taylor & Ward, 1994).

EXPORT-ORIENTED DEVELOPMENT: INTERNATIONAL CAPITAL, THE MALAYSIAN STATE, AND LABOR

The traditional economistic explanation for the diffusion of capital to the Third World based on the concept of "comparative advantage" of existing productive factors is contradicted by a host of evidence. Although availability of low wages and low skills are certainly important to investment decisions, it is state behavior, including the legislation and enforcement of antiunion policies and the promotion of industrial peace, that is ultimately the most crucial element in the courting of foreign investment - apart from the market-seeking inclinations of capital itself. Were this not the case, the extremely low wages and other valuable resources in much of sub-Saharan Africa, for example, would have made those countries the principal sites of direct foreign investment (DFI), and, of course, this is hardly the reality. Indeed, almost half of the DFI in the Third World is concentrated in some 20 countries, only a few of which have shown significant growth since the export-oriented industrialization drive began some 25 years ago. By 1972, five NICs held 54% of all Third World manufacturing assets (Cohen, 1987), and by the 1990s, the four ANICs produced over half the Third World's industrial exports (Douglass, 1991).

State-directed incentives for DFI may include full repatriation of TNC profits, tariff reductions or exemptions on imported parts, patent protection, accelerated depreciation, income and capital gains tax holidays (usually for 5 to 10 years), guaranteed currency exchange rates, access to local financing, minimum state regulatory enforcement, and cheap (no regulated minimum) labor costs. Even after the termination of pioneer status privileges, transnationals in Malaysia were still given the benefit of an investment tax allowance, which amounted to the same thing. Other important associated incentives offered by the state include surveillance and control of labor organizers and other policies to assure a docile workforce reticent to resist low pay, poor benefits, little job security, and substandard safety, environmental, and health conditions (Rasiah, 1994b; Stockton, 1991).

As an important actor in establishing platforms for offshore electronics and in setting the conditions of production and work, the Mahathir government through UMNO has been a virtual holding company for many of the largest domestically controlled and joint venture industries in the country (*Far Eastern Economic Review*, 1988; Seward, 1988). Catering to investment objectives, nothing in Malaysia's official development (New Economic Policy) literature explicitly suggests that electronics workers were intended for inclusion in redistribution efforts of the state, except in fostering higher quotas in factory work for majority ethnic Malays and other supposedly indigenous peoples (*bumiputras*) vis-a-vis the more affluent Chinese minority. The concept of "national development" is effectively founded on privileged property rights of foreign capital and a small class of commercial and wealthy Malaysian subalterns committed to subsidizing transnational investment, with those outside the inner circle of policy makers left to fend for themselves. State incentives to foreign capital (including 100% equity participation), expected to run out in the 1980s, was extended into the 1990s (Salih & Young, 1989).

NIDL AND STATE SOCIAL ENGINEERING: REINTEGRATION OF PENANG

The modern history of Penang goes back to the island's origins as a rice cultivation and fishing economy. It was the British colonial administration that transformed the island into a free port (1786) of traders, merchants, and spice growers. In the early 1900s, Penang was a commercial center for tin, rubber, and agricultural commodities, but

Singapore actually served as Britain's main entrepot at the time, leading to Penang's decline. In the 1960s (1963-66), Indonesia's "confrontation" (*Konfrontasi*) with Malaysia reduced the latter's trade with Burma, as Thailand cut back its trade with the island. By 1969, the year of Malaysia's communal riots, the heavily Chinese-populated island had its free port status removed.

In the 1980s, the largest semiconductor assembly center in the world was located on the island of Penang (Pulau Pinang in the Malaysian language) and transformed the economy. Of Malaysia's total manufactured exports in 1988, electronics overall contributed 56% (US$ 5.5 billion) over a quarter of its 1989 GDP, a remarkable transformation of a recently rural-based economy (*Global Electronics*, 1989; Stockton, 1991). Semiconductors still constitute 40 percent of Malaysian electronics production, but computer parts assembly and consumer electronics have grown much faster. The aggregate sales of Malaysian electronics in 1994 was $15.5 billion (Bangsberg, 1995).

More than half the value (56%) of Malaysia's electronics components exports in 1992 came from U.S. companies, with their 41,000 employees, the country's largest employer in the industry (Ismail, 1995). The industry's workers are still primarily involved in assembly of integrated circuits ("chips") but are shifting increasingly to consumer and industrial electronics and to more advanced technologies, including disk drives, which already employs 15% of the total Malaysian electronics workforce (comprising 169,545 workers as of 1994 and 110,375 in 1990, a 54% increase; Ghazali, 1994). Japanese firms, the biggest investors in the industry as of 1996, have been involved in manufacturing higher end consumer electronics and electrical appliances. The biggest of these firms include Matsushita (color televisions), Sharp (color televisions), and Sony (color televisions, video cassettes, audio products, chips, and micro-floppy disk drives). Malaysia is Japan's primary source of imported televisions and second largest source of video-cassette recorders. Nearly all the components needed for assembly in Malaysia are imported (Kaur, 1995)

At the end of 1994, Penang had 695 factories, of which 156 were in the electronics and electrical sector, an 8.3% increase over the previous year, employing 101,838 workers (Penang Development Corporation, 1995). Electronics/electrical companies had almost 58% of Penang's total manufacturing employment (others include textiles and garments, basic metals, fabricated metals, and machinery). As one indication of the confidence that foreign electronics corporations have in Malaysia as a base of operations, the Harris Corporation (U.S.), which manufactures telecommunications, broadcast, and microwave radio equipment, declared in 1995 its intentions of making the country its research and development center of operations in this sector, and of investing more than US$100 million between 1995 and 2000 (Ong-Yeoh, 1995).

Although high costs of living are driving many Malaysian workers and managers out of Penang, it remains, with its 12 industrial estates, the preeminent export production area in Malaysia, with about half of the country's electronics workforce. Foreign investment in these 12 estates in 1994 was valued at US$1.8 billion, with electronics exports at US$7.73 billion. Electronics forms 40% of total Malaysian exports and 80% of Penang's GDP (Harun, 1995).

MALAYSIA'S ELECTRONIC WORKFORCE: ISSUES AND PROSPECTS

Ten years ago, Malaysia's niche in the electronics international division of labor was in semiconductors, but since then, transnationals have introduced to Penang higher stages of technology with the manufacture of computer disk drives, devices that until recently were largely produced in neighboring Singapore—and at higher wage rates. Leaders in Malaysia's electronics industry hope by the late 1990s to bring computer storage manufacturing and the full range of computer production to the country. Penang and other manufacturing regions in Malaysia currently produce many of the world television sets, cellular phones, audio equipment, and many other artifacts of the information age.

Most of the information and communications end-use equipment that Malaysian (and other Third World) workers help to produce will never be available to them, and Western end-users by and large have no idea about the conditions of labor embodied in these commodities. There is virtually no education about labor in the United States, and the country's major socializing institutions—the public schools, universities, the mass media, political party organizations, even trade unions—do little to raise public consciousness about transnational corporate activities and treatment of workers in Third World countries. In Malaysia, the range of political freedom to question these relations of power in the workplace are considerably more constrained, and ideological apparatuses such as mass media are more directly controlled by the ruling political elite.

Among the early U.S. electronics transnationals in Malaysia were National Semiconductor, Intel, Texas Instruments, Motorola, RCA, Advanced Micro Devices, and Hewlett-Packard. The largest electronics factories in Penang, measured by capital reinvestment in 1994, were Intel (U.S.—RM 328 million), Komag (U.S.—RM 135 million), Hitachi (Japan—RM 111 million), Hewlett-Packard (U.S.—RM 77 million), Sony Electronics (Japan), ASE (RM 55 million), Conner Peripherals (U.S.—

RM43 million) and Siemens (Germany—RM 39 million; Penang Development Corporation, 1995). Electronics was from the outset almost entirely in the hands of foreigners, who had a 91% share of ownership in 1993 (Rasiah, 1995b), although local manufacturers, almost all Chinese and many with past work experience in the TNCs they service, have created a small niche for themselves as suppliers of machine tools to the transnational firms, and some have used their TNC connections to open export markets outside of Malaysia, mainly to Singapore. Domestic manufacturing firms increased their export/output ratio from 0.4% in 1983 to 31.6% in 1990 (Rasiah, 1994b).

Of the country's electronics workers, more than half are employed by U.S. companies, mostly in the assembly of integrated circuits, in 19 semiconductor plants as of the early 1990s (Narayanan & Rasiah, 1992). A number of Japanese firms have invested and planned expanded investment in higher end consumer electronics and electrical appliance production, including Matsushita (color televisions), Sharp (color televisions), and Sony (color televisions, video cassettes, audio products, chips, and micro-floppy disk drives). Two of the Asian "newly industrializing countries" (ANICs)—Taiwan and Hong Kong—also had established electronics operations in Malaysia (Global Electronics, 1989). Nearly all the components needed for assembly in Malaysia are imported (Salih & Young, 1989).

The Role of the State

In the Third World, the state is central in generally creating conditions favorable to direct foreign investment, reproducing world market structures within the domestic economy, and delivering workers to transnational corporations, even as it reduces the level of its own direct ownership of domestic enterprises. When Malaysia's Mahathir government resorted to state repression in the form of arrest and torture of labor organizers, the low wages and substandard working conditions such practices fostered had the effect of discounting investment costs to transnational capital. When the AFL-CIO unsuccessfully pressured the U.S. Congress in the late 1980s to end Malaysia's GSP status, the Malaysian government offered tactical policy concessions on the right to organize electronics workers in order to protect both local export-oriented clientelist bourgeoisie and their transnational partners. Mahathir, however, has consistently opposed a national electronics union and recognizes none of the eight in-house (enterprise) unions, even though most of them are little more than social organizations (Vatikiotis, 1992).

The state's key role in the formation of free trade zones in Malaysia has privileged TNC investment but led to little technology transfer or new domestic value added enterprise. With unrestricted import inputs, TNCs have little incentive to use local content or create domestic research and development facilities, although competition among them has resulted in increased local sourcing of machine tools. The Malaysian state, as yet, has not followed the South Korean approach of subsidizing and protecting domestic corporations, for example, the giant *chaebols* such as Daewoo, Samsung, and Hyundai, or forcing local content restrictions on foreign transnational corporations, except in very limited and weakly enforced cases. Whatever local sourcing exists has more to do with transnational incentives based on changing international conditions of production (Rasiah, 1994b). In Malaysia such incentives largely favor Chinese firms and are inconsistent with official Malay-first (*bumiputra*) policies.

The role of the state as an agent in the motion of transnational capital has in general been understated and understudied. The historical disposition of the state toward labor and capital varies with the social (class, ethnic, linguistic, etc.) composition of the state, primarily in terms of its class orientation and patron-client constituencies. Despite prolonged "Emergency" rule (1948-60) by the British, Malaysia followed a relatively peaceful transition to self-rule in the late 1950s and early 1960s and absorbed most of the political and economic institutions of the British colonial regime, later adopting the ethnically protectionist *bumiputra* policy.

In practice, the dominant political party, UMNO, serves as a bureaucratic-capitalist association of politicians, political organizations, communitarian parties (overwhelmingly Malay but including some Indians and Chinese), and major domestic and joint venture enterprises, including business partnerships in electrical and electronics industries. UMNO has share control over the dominant media conglomerate, the Fleet Group (the *New Straits Times* and other leading newspapers and newsprint company), two other Malay language newspapers, Utusan Malaysia and Utusan Melayu, a Chinese language newspaper, MetroVision (over-the-air) television, TV3 (a private television network), and many other financial, service, media, and industrial enterprises. MegaTV (subscription television) is financially controlled by the Ministry of Information (Keenan, 1995).

Given its extensive direct linkages to domestic and international capital, the Malaysian state can not be regarded as even relatively autonomous in its relationship to the private sector. UMNO's linkages to domestic and foreign capital leave its leadership little choice but to cooperate with transnational industrial, energy, finance, mining, and agricultural interests, which are the pillars of the country's export economy platform. The stress of adapting to an integrated and competitive system of capital formation imposes authoritarian tendencies on the state in dealing with diverse political currents and with politically conscious trade unions. A number of social scientists have noted the relationship of rising industrialism, in particular, export-oriented industrialization, in the Third World to the implementation of state repression and strict limits on political expression (Deyo, 1987; O'Donnell, 1973). The Malaysian governing apparatus has little preparation and few resources to manage the complexities of power relations with international capital, which is backed by other political and economic institutions in the core industrial states. In high-tech areas such as electronics and communications, the Malaysian state is coopted by the developmentalist logic of international capital to an even greater extent.

Although the Malaysian state has, in principle, some control over the conditions embodied in technical agreements with foreign corporations such as in technology and investment level, local equity and content participation, R&D, patents, trademarks, human resource training requirements, and so on, of which electronics constituted 22% of the total number from 1975-1993, in practice these considerations are taken up only at the approval stage, with little or no followup. In 1992, 84% of all R&D in Malaysia went to electrical/electronics industries, of which more than 91% was undertaken by TNCs. It is argued that the government does not have the technical capacity to evaluate and implement its agreements with foreign corporations, lacking mechanisms for appraising and monitoring them. The results of the agreements are rarely if ever studied, although a government-industry partnership, the Malaysian Technology Development Corporation, was created in 1992 to encourage advanced local participation in R&D (Rasiah, 1995b). If the study of technical agreements with foreign capital is regarded as difficult, officially sanctioned research on the relationship of foreign investment to the domestic labor force is out of the question.

The Makeup of the Workforce

In the late 1970s, transnational corporations began to respond to concerns about productivity, costs, and quality control by shifting to more fully automated systems of production. The expectation of some was that with greater automation, electronics TNCs, less dependent on labor-intensive processes, would pack up and leave countries like Malaysia. The electronics TNCs did not leave and, in fact, expanded production. Nor did they transform the composition of work from an unskilled labor force into a skilled one. Although automation has expanded, the composition of electronics and electrical employment in Penang—48%—is still categorized as unskilled, with 26% skilled (Penang Development Corporation, 1995; see figure 6.1).

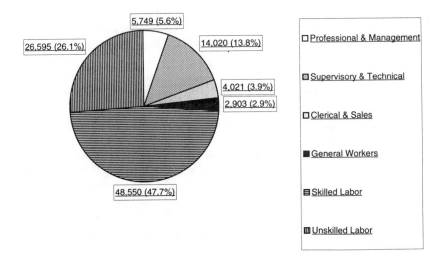

Figure 6.1. Electrical/electronics employment in Penang, December 31, 1994 (156) factories
Source: Penang Development Corporation, 1995, p. 6.

In adding more technology to the mix with labor, electronics corporations were able to extract greater productivity with small increases in wages. In one semiconductor firm in Penang, a foreign corporation was able to shift from one die-attach machine per worker in 1972 to eight in 1990. Moving away from Fordist and into "just-in-time" (JIT) production methods, Malaysian workers are required to rotate between multiple tasks when work is completed in one. JIT reduces the required number of stocks, equipment, and employees, often sharply cutting back the number of workers per plant (Narayanan & Rasiah, 1992).

The reduced need for labor per plant, therefore, tends to offset the expansion of the industry as a whole. Workers cannot use the growth of electronics, even with a 40% market growth for semiconductors in 1995 (Markoff, 1996), as a lever for better bargaining conditions. The protected status of foreign corporations in the industry makes them particularly invulnerable to efforts of workers to unionize (far more so than in local companies) and improve material and physical working conditions. Foreign managers may, as in the case of U.S. corporations, use informal management practices over production workers (operating on a first-name basis, joining them in out-of-work sports, sitting at the same table) to create a false aura of egalitarianism. Japanese corporations tend to be more tolerant of at least in-house unions but are more formal in their management relations with labor (O'Brien, 1994).

Gender

One of the central characteristics of transnational enterprises in the Third World is their heavy reliance on female factory workers, globally between 80% and 90% in offshore manufacturing, but confined almost entirely to the labor-intensive areas of production. In supervisory and management areas, the percentage of female employment drops off precipitously. This pattern is also found in the worldwide electronics industry. In Penang's electronics factories, women hold 72% of the jobs (a considerable higher percentage if supervisory and management positions are not counted), an overall decline from the early 1980s when the average was 80% or higher (Penang Development Corporation, 1995). In East and Southeast Asia, in general, women make up the largest rates of female factory employment in the world, nearly half the manufacturing workforce in Malaysia, and more than one fourth of the region's total manufacturing workforce (United Nations, cited in Cammack, Pool & Tordoff, 1993).

Electronics employers have traditionally explained that this preference is based on special physical and psychological attributes of females: finger dexterity, especially in such tasks as die-attachment; wire bonding, and printed circuit board insertions, the ability to operate small mechanical objects; and endurance for tedious work. Critical observers argue that such claims are mere obfuscations for cultural and political forms of exploitation, and that the actual reasons relate to the productivity advantages of low wages (Third World females earn on average a little more than half their male counterparts), less union interference (females tend to organize less frequently and less militantly than males), and subordination (females tend to put up less resistance to management demands). When Malaysian women do receive average unskilled wages (US$130/month in Kuala Lumpur), they still remain part of the official 17.1% (1990) of the population living below the poverty line—in a country where the GDP per capita in 1990 was US$2,460, one of the most income-polarized economies in Asia. (The top 1% of shareholders held more than 32% of the country's corporate shares; the top 5% had 63% in 1983—Mehmet, 1986).

Transnationals and the Malaysian state collaborate on a fundamental level of labor exploitation in the patriarchal employment of female workers, who are compensated at 10% to 12% percent of their Western counterparts (69% of Malaysian males) but are expected to carry a 50% heavier workload (Momsen, 1991). Women are valued as having particular utility in Malaysia's electronics (and textile) industry. Throughout the Third World's export-processing zones (EZPs), women in fact constitute about 90% of the labor force (but only 20% of employment within Third World TNC subsidiaries overall). Malaysian women in manufacturing, subjected to "triple exploitation" (as workers, females, and non-whites) are seen by TNCs and their local partners as more physically and psychologically adaptable to the rigors and monotony of unskilled and semi-skilled labor. A personnel administrator at Intel, claimed, "We hire girls [sic] because they have less energy, are more disciplined and are easier to control" (quoted in Levidow, 1991, p.106).

Penang's largely female electronics workforce, around 80% until 1980, is recruited primarily from the rural areas of the island and from nearby states. Malaysian culture is strongly male-dominated, and the resurgence of conservative Islamic traditions since the early 1980s (influenced by the coming to power of Khomeini in Iran in 1979) has added another layer of patriarchal supervision in the lives of Malay women (non-Malays and non-Muslims are exempt from formal Islamic practices). As in the Middle East, Malay women are expected to cover their heads, legs, and arms in deference to Koranic orthodoxy. Within the confines of electronics factories, women are also expected to demonstrate subordination to male authority figures. As a result, the transnational corporations in Penang and elsewhere in Malaysia have come to expect a well-disciplined, communally divided workforce and labor peace at wages about one eighth or one-tenth developed countries' standards and with high levels of productivity.

The conservative religious resurgence in Malaysia, with its traditional rejection of secular liberal values, tends to support labor's acquiescence in state control, at least among ethnic Malays. Religious orthodoxy and traditional patriarchy, though considerably less confining compared to present-day Saudi Arabia or Afghanistan, have imposed moral guidelines on women to shun more independent self-expression, dress style, and social practices, discouraging initiatives toward meeting with male union organizers or taking up factory organizing on their own. This has not, however, altogether prevented resistance. In 1988, to cite just one example, female workers at a Japanese company, Mitsumi Malaysia, successfully went on strike to force better wages and work-shift schedules (Kimori, 1989).

It is notable, though, that labor unions have had little success at organizing in this sector, whether in Malaysia, the Third World, or the United States. This can be explained to some extent because of cultural factors placing heavy demands on female conformity, especially in a context like Malaysia's in which extended families in the rural areas often depend on the supplementary income earned by young, unmarried women in the export processing zones, often to help pay for the education of their male siblings (Lin, 1987). The high-tech aspects of electronics production is another cultural factor discouraging female mobilization because the computer industry as a whole still remains a bastion of male domination, from systems design to video game users

Earlier studies described how management in Penang and elsewhere in Southeast Asia often attempted to pacify the underpaid female workforce with non-wage rewards and diversions, such as beauty contests, singing competitions, and individual awards for productivity (Grossman, 1979). Islamic codes eventually pressured managers to eliminate the beauty contests, but companies still use other forms of competition (prizes, dinners, sometimes travel) to induce "company spirit" and higher levels of productivity. Women are also pressured by their male employers to attend company picnics, dress-up affairs, and lectures to breed corporate loyalty and stimulate competition among different production lines (Chan Lian Heng, personal communication, August 28, 1995). Within the context of a restrictive patriarchal culture, it is also true that many women see factory work as "liberating" them from direct parental control and enabling them to bargain even small savings for more freedom in their home villages, for example, by making them more financially independent and allowing them to choose their own marriage partners (Lin, 1987).

Factory life is also proletarianizing women, who already make up nearly 50% of the total Malaysian workforce. As Third World countries become increasingly industrialized, they become more exposed to the specific inequities of world capitalist integration. It becomes more difficult to sustain a traditional patriarchal domination of women when they are needed to supplement family income, whether in the formal or informal sector. This occurs in the industrial zones and also in those agricultural areas pushed into cash crop and export-oriented production. The provincialism of the past inevitably over time leads to new exposure to the secular and material world developing around them and to new emerging values regarding the status of women. The growing impoverishment of their Western female proletarian counterparts may help to unify them along lines of both class and gender.

Ethnicity: Bumiputra/Non-Bumiputra Issues

Since coming to power in 1981, Mahathir Mohamad has used ethnic
bumiputra politics to strengthen the UMNO elites and to prompt Malay
worker and peasant loyalty toward the state. Enshrining the state as the
defender of Malay and other "indigenous" peoples' interests, it is
considered sabotage for workers to militantly press their demands in the
workplace, and TNCs have not challenged the discriminatory policies
any more than they support affirmative action quotas for ethnic
minorities in the United States. Economic development putatively
directed on behalf of the whole "indigenous" population is embodied in
the provisions of NEP (since 1990 renamed the National Development
Policy or NDP), which has helped protect the traditional economy of
Malay small farmers and landed aristocrats from control by Chinese
capital (Limqueco, McFarlane, & Odhnoff, 1989). (At the same time, only
a small percentage of Malays have ancestral connections to the land that
date prior to the 19th century, and many with *bumiputra* status are, in
fact, recent immigrants from Indonesia.) One critic of the Policy
commented: "Bumiputra elites who have benefited from NEP trusteeship
are small, powerful and influential people who gain through collusion,
transaction costs and other forms of non-competitive gains . . . [and] are
the ones in the best position of access to vital information regarding
contracts, tenders and other investment opportunities" (Kua, 1992, p. 58).

The ethnic breakdown of the Malaysian population is as follows:
59% Malay and other "indigenous," 32% Chinese, and 9% Indian (and
other). Penang has a very different ethnic makeup: 52.9% Chinese, 34.5%
Malay, and 11.5% Indian (1.1% other). One area in which the NEP (and
NDP) has apparently altered Malaysian demographics is in the ethnic
composition of the working class. Malay proletarianization has grown
from 16.7% in 1957 to 34.2% in 1970 and 48.5% in 1990 (Crouch, 1993).
Much of this growth, however, represents greatly increased female
participation in the workforce. Although working-class Malay women
have made many advances in industry, *bumiputra* policies have not
altered their concentration in the lowest paid, labor-intensive assembly
operations of food, textiles, and electronics.

In addition, although the NEP did help to create more industrial
work opportunities for Malays as a whole (62% of the electronics
workforce being Malay by 1986), only a small percentage moved up to
management positions (Narayanan & Rasiah, 1992). Malay women
dominate the electronics industry assembly jobs, Chinese women the
clerical and supervisory positions, and Indian men the union organizer
roles. On the whole, communal allegiances in Malaysia have had the
effect of impeding a unified class identification and allowing the

government to implement authoritarian controls to destroy the labor movement (Crouch, 1993).

Changing Composition of Work

The nature of the work process in electronic components has been transformed since its introduction to Asia in the late 1960s. Automation and statistical process control techniques are now used extensively in the industry. In wire bonding, for example, workers are expected to operate from 4 to 12 machines simultaneously, monitor inventory and defects, and take responsibility for material supplies and minor machine repairs: "Such changes are an outcome of flexible production organizations, implemented for the purpose of achieving ever-increasing productivity, and they have resulted in stretching work boundaries both horizontally and vertically" (Rasiah, 1994a, p. 21). Between 1985 and 1987 alone, the ratio of capital to labor increased from 21:6 to 19:2 (Narayanan & Rasiah, 1992), although consumer electronics continues to be considerably more labor-intensive than the components (semiconductors) subsector.

In recent years, Malaysia has followed the Japanese industrial pattern of restructuring work along the model of "quality control circles" (QCC). This approach attempts to boost productivity through "consensual" worker agreements, which includes their participating in discussions of production goals, organizing tasks in smaller work groups, and having them learn and engage in multiple tasks to vary daily work assignments. QCCs are designed to discourage union inclinations, a strategy consistent with the anti-labor policies of the Malaysian state (Jomo, 1990: 218). However, unlike Japan and despite Mahathir's "Look East" rhetoric, Malaysia does not assure lifetime employment, nor anything near the wages and benefits of their Japanese counterparts.

Flexible production techniques were developed for the purpose of reducing inventories, embodied technical costs, defects, and overhead. A related concept, "just-in-time" production methods, developed for large-scale output by Toyota to customize output and to perfect buyer-seller transactions toward zero inventory, further tailors and Taylorizes the subdivision of work (the managed control of shopfloor labor, time, mechanical motions, and output) to the demands of corporate contractors and subcontractors. All the major electronics transnationals in Malaysia have introduced flexible and JIT production systems, resulting in falling numbers of workers per plant, while they continue to shift to computer-controlled automated production (Rasiah, 1994b).

Automation is forcing both a deskilling trend in certain production areas and higher level skills in others, generally intensifying the work pressure and, with few exceptions, providing no new skills transferable beyond the plant. National Semiconductor, for example, automated wafer cutting, bonding, molding, and testing operations in 1989. Assembly work is being substituted by higher end production processes and by low-paid, white-collar employment. According to some observers, in component manufacture, but less so in consumer electronics, line workers' routinized work processes have begun to shift toward more statistical, problem-solving, and conceptual skill applications (Rasiah, 1994a; Salih & Young, 1989).

Automation in microelectronics to a certain extent has already forced Malaysia to shift to less labor-intensive areas and to develop forward linkages to other microelectronic and to higher end communication technology production. A surplus regional labor pool from among Malaysia's neighboring countries is another factor. Low-wage competition from poorer entrants to the labor market such as China and India can be expected to further diminish the appeal of Malaysia's assembly workforce. Having a combined population of some 2.2 billion people in 1996, China and India will be able to make compelling bids for electronics investments, with extremely low cost labor, a well-trained workforce, and extraordinary in-country market opportunities for TNCs.

Labor Shortages

The expansion of the electronics industry worldwide has contributed to Malaysia's general economic growth. At the current rate of expansion in Malaysian electronics, it is anticipated that 320,000 new jobs will have to be added—120,000 in Penang alone and 41,845 for the island for the years 1995-1997. With a population of 1.7 million (in 1993), this will mean rapid immigration, with many, most probably, from other countries such as Indonesia (Sumatra), the Philippines, Bangladesh, southern Thailand, and perhaps Vietnam. In Malaysia as a whole, there are already more than 1 million foreign workers in a total workforce of 7.5 million, which may continue to expand in the next few years. This will undoubtedly create severe problems and tensions in a country with already delicate ethnic relations among the dominant groups—Malays, Chinese, and Indians (Harun, 1995).

However, more than half (57%) of this new demand for electronics workers is expected to come under the "skilled" category, which, together with a currently tight labor market, should drive up average wages in the industry. At the same time, under such circumstances, it will be difficult to force companies to achieve hiring quotas for *bumiputras*. Also, relative reductions in the hiring of unskilled categories will mainly affect Malay women and further undermine the economic restructuring goals of the ruling UMNO. Chinese males tend to have more formal preparation for the technical and managerial levels of employment, whereas Malay professional and technical employment is concentrated in the fields of teaching and nursing (Narayanan & Rasiah, 1992). Hence, technical restructuring by TNCs is likely to have powerful class, ethnic, and gender repercussions on Malaysian society.

Conditions of Work

It is often argued by their defenders that TNCs provide better overall working conditions and wages than domestically owned firms. It also has been argued, even by Marxists (e.g., Warren, 1980), that, led by TNC-controlled firms, a form of Third World capitalism and proletariat is emerging that leads to the organization and articulation of working-class demands. As Michael Burawoy (1985) notes, however, such claims are not well grounded empirically and usually do not examine the nature of the production process. It can also be said that even when TNCs do contribute to enclaves of industrial capital and labor in the Third World, they also create contradictory outcomes, including particular forms of state repression, relative and absolute labor exploitation, and wider gaps in social income distribution (Berberoglu, 1992).

It has often been argued by critics that TNCs look to the Third World for investment mainly to exploit the extremely low wages compared to the OECD countries. While low wages and poor benefits continue to be powerful inducements for relocating production, they are not the only factors explaining TNC investment objectives. Other important factors include political stability, good physical infrastructure (e.g., roads, airports, seaports, telecommunications, reliable electrical power, etc.), financial incentives, a trainable labor force, friendly governments, little bureaucratic interference, and proximity to markets. In the case of Malaysia, wages are very "competitive," especially at the skilled and semiskilled levels, but the other factors are critical as well, particularly when compared to neighboring countries (the Philippines, Vietnam, Indonesia, China, and Thailand).

In recent years, Malaysia has shifted from providing a largely unskilled workforce involved mainly in assembly operations to a more technology-intensive work environment in which a higher ratio of skilled employment is needed. With fewer assembly-line jobs, the average wage has gone up, but not nearly as high as productivity. Hence, with a still very low-wage workforce, a stable political system that welcomes foreign investment with attractive incentives, and mostly nonunionized factories and compliant workers, the electronics TNCs have little reason to leave. As Narayanan and Rashiah (1992) note, "cheap *unskilled* labour has merely been replaced by the need for cheap *skilled* labour, to operate more sophisticated processes. Parent sites are not sources of cheap labour—skilled or unskilled" (p. 87; emphasis in original).

Hence, electronics workers are experiencing limited growth in wages. However, in an industry as changing, inherently unstable, and unpredictable as electronics production, Malaysia's workers can neither count on long-term job security nor opportunities for job advancement. Competition between chip makers in the U.S. and Japan and the high cost of R&D in a rapidly changing technological field require TNCs to keep down labor costs, and in this they enjoy the support of the Malaysian state. Until recently, high levels of worker turnover in a labor-surplus market contributed to suppressing wages, while enabling companies to replace exhausted workers and to discourage worker organizations. The Malaysian press now speaks of a labor shortage in the industry. This raises the question of whether Malaysian labor will be able to demand higher wages and benefits, competing more actively in higher end production with Singapore (disk drives) and India (software engineering), or whether the state will more actively recruit workers from abroad who are willing to work for less than their Malaysian counterparts.

Worker Discipline and Wages

Taylorist methods of production require strict discipline of worker behavior in the factory. In Malaysia, the traditional patriarchal culture of authority encourages foreign investors to expect little sabotage or protest against the demands and pace of work. This means in practice, among other management demands, that no talk is permitted during work periods, and bathroom breaks are few and of short duration (15-minute breaks once every three hours). One observer noted how one woman brought respite to her tired fellow workers on night shift by short-circuiting a machine that was the immediate source of their exhaustion (Tan, 1986.). In recent years, women have adopted Islamic injunctions to force companies to provide areas and breaks for daytime prayer.

Originally prohibited by legislation intended to protect female workers, rotating shift work and permanent night work were later exempted from regulation as automated equipment came into use (Lin, 1987). Shifts run from 7 a.m. to 3 p.m., 3 p.m. to 11 p.m., and 11 p.m. to 7 a.m. Rotating shifts causes workers disruptions in their living routines. Many complain of skipping meals or eating on the run, which often leads to severe gastritis. Married women often prefer to work the night shift in order to spend more time with their families in the daytime, but struggle to stay awake on the job. Weekends are given off only once every 6 weeks. Others prefer shift work because of the extra pay—RM1.50 (US$0.60) for afternoons and RM2.50 (US$1) for night work. Without shift work, normal pay for electronics work in the early 1990s was RM335 (US$134) per month (Chee & Subramaniam, 1994), in 1995, between RM 380 and RM 400 (US$152-US$160) was standard for entrants in the Kuala Lumpur region (Chee Heng Leng, personal communication, September 4, 1995). Another study found that one U.S. electronics firm paid as high an average wage as RM690 (US$276) per month (Rasiah, 1995a), which is still only a small fraction of their poorly paid counterparts in Silicon Valley.

In Penang, according to official sources, basic wages in the electronics/electrical industry (mid-1994) were RM12.5/day (US$5—unskilled), RM20/day (US$8—skilled), RM750/month (US$300—technician), RM2200/month (US$880—engineer; Penang Development Corporation, 1995). Wages in electronics are on average higher than in the garments industry but lower than in manufacturing overall, although wages in the semiconductor subsector are higher on average than in manufacturing (Narayanan & Rasiah, 1992). The influx of workers from rural areas has in the past helped to keep down wages. Future industry growth may indeed rely more on Indonesians, Filipinos, and Thais more desperate for low wage work.

Occupational Safety and Health

Aside from low wages, one of biggest "rents" paid by electronics workers for the privilege of being exploited is the damage done to their health. Workers interviewed by several Malaysian scholars have reported high rates of lung cancer deaths among otherwise healthy women working in the mold rooms where inhalation of toxic particles is common. Attracted by the relative material benefits—an allowance of RM10.00 (US$4) added in 1991 on top of the then daily wage of RM7.50 (US$3)—female workers are willing to take on such risky jobs, and also because there are no better income alternatives (Chee & Subramaniam, 1994). Risk assessments are thus contingent on available information but also the perception of next best options. Chee and Subramanaim (1994) note:

> The notion that individuals are solely responsible for their own safety and health is one that dominates in our [Malaysian] society; hence, the workers themselves sometimes internalize it. This creates a dilemma for them, because it often means sacrificing their well-being for the sake of their jobs. One sees different responses to this dilemma; some rationalize philosophically, while others try to protect themselves in whatever way possible. (p. 100)

One of the worst jobs, one that is being phased out of some operations through automation, is the use of microscopes ("scopes") for such applications as die bonding, a work hazard strongly associated in several studies with deteriorating eyesight. Workers are made to feel that electronics components are given more value and respect than the workers themselves. One worker described her experience this way:

> When I use the scope, I cannot eat, my head aches, I feel like scolding people only, tension only, don't feel like talking to people. I feel the pain every day. Every day, I use the scope. Night shift is the worst. Inside the scope, they have put a light. When I look at it, I feel as though the light goes into my head. . . . The supervisor will scold if we do not switch on the light in the scope. He says the light is to help to see the rejects clearly. (quoted in Chee & Subramaniam, 1994, p. 91-92)

One of the ways by which employers control women workers is by restricting access to information. The health risks from using microscopes and inhaling toxic solvents, acids, and sometimes radioactive gas arising from this type of work are not communicated by companies to their employees. Illnesses are not covered by employers or insurance companies and often go untreated. Shopfloor workers usually wear overalls and work in air conditioned environments, but these conditions are intended to protect the components, not the employees (Consumers Association of Penang, 1994).

In Malaysia, semiconductor workers have three times the rate of occupational-related illnesses of other manufacturing sectors. The wafer etching process used in Malaysian electronics plants is said to be responsible for a 39% rate of miscarriage among pregnant workers. The common use of trichloroethylene, methyl ethyl ketone, xylene, acetone, solder flux, sulfuric acid, and hydrochloric acid are associated with cancer and reproductive problems (Lin, 1987; Stockton, 1991).

An electronics worker at the Seagate (U.S.) electronics company in Penang wrote to the Consumer Association of Penang (CAP) in July 1990, urging that the human rights organization investigate conditions at the plant:

> I am an operator working in Seagate Industries [at the] Bayan Lepas FTZ, Penang. We work on disk drive peripherals. These products require cleaning. The solvents used for cleaning are "TP 35," "TMS" and "TF." These are alcohol based solvents and "freons" which is [sic] harmful to our lungs when inhaled. Many operators are working in "freon" exposed environment[s] thus inhaling evaporated "freon" in "clean rooms," endangering their health. The company has done nothing to assure our health safety. The only safety measure taken is wearing a face mask which is insufficient.

Another electronics employee, in this case a process engineer, also complained in a letter to CAP of the conditions at National Semiconductor (U.S.) at an industrial estate near Seremban (south of Kuala Lumpur) in May 1994:

Each time I walked past the frame wash room, I could get a strong smell of TCE (tri-chloro-ethylene) and the girls working inside sometimes used to complain of feeling dizzy and sick due to the fumes. Whenever the issue of installing a gas duct was raised with the plant manager, who was an American, his reply was that a recession was coming and that the management did not want to spend any money.

As the Consumers Association of Penang (1994) noted:

Production workers are not the only ones at risk. The factories also engage young male workers on a contract basis to pour out chemicals from big drums into smaller containers for use in factory operations. These workers know little about the health hazards involved and are sometimes not even provided protective equipment.

Clearly, these forced rents in the forms of low wages, poor benefits, and weak health and safety protection are embedded in the immediate conditions of bargaining that now exist. Shortsighted policies on the part of TNCs and the state relate, in turn, to the question of worker empowerment.

Unions

During the struggle for national independence, Malayan workers (before the Malaysian Federation was conceived) organized the first trade unions in the 1920s and 1930s under conditions of colonial British repression. It was not until World War II that Britain legalized trade unions. By the end of the war, Britain returned to policies of labor repression, especially against the General Labour Union, its successor the Pan-Malayan Federation of Trade Unions, and other unions that were part of the anticolonial movement. Britain used various instruments to deter the unions, including police, private security guards, thugs, civil and industrial courts, the Ministry of Labour and Manpower, the Registrar of Trade Unions, and detention centers. Led by UMNO, the new government on which the British would confer partial independence in 1957 (Malaysia—merging Malay with Singapore, Sabah, and Sarawak—was formed under British supervision in 1963, with Singapore withdrawing in 1965) retained many of these repressive anti-union laws and institutions into the 1990s (INSAN, 1987).

With the establishment of the new Federation, UMNO has used a variety of means to control unions. In 1959, it reinvoked a Trade Union Ordinance, drawn largely from draconian British era Emergency (1948-1960) decrees (that had tens of thousands of "anti-imperialists," as the British authorities themselves labeled them, sent to their death to jail or deported—mainly to Chiang Kai-Shek's China). In 1965, compulsory arbitration was imposed, which limited the right to strike in "essential services" (which could include almost any industry). An Industrial Relations Act was introduced in 1967 that provided "protection of pioneer industries during the initial years of its establishment against any unreasonable demands from a trade union" and waived other protections for workers. In October 1969, amendments to this Act were passed that included the right of management to dismiss workers, force transfers, and use other methods to bar trade union initiatives, and a 1980 amendment to the Act banned unions in these industries indefinitely (Wangel, 1988).

In the 1970s, fearful that labor might form a party similar to the Labour Party in Britain, the government passed new legislation that forbade union officials from holding office, fraternizing with politicians, or using the strike weapon against nonrecognition. If these laws were not enough, the 1960 Internal Security Act (ISA), a carryover from the colonial era, and the 1972 Official Secrets Act (OSA), forbidding unauthorized publication of information in the hands of the government no matter how trivial or widely known, have been used, as in 1987, to detain without trial anyone seen as endangering the security of the state (as well as intra-UMNO challenges to Mahathir's control that occurred that year). Under the ISA, many union leaders and oppositionists affiliated with the Labour Party and other political groupings were detained in the early 1970s, thereby effectively killing off the Party structure, and again, in 1987, some 106 labor and political activists were detained without trial to discourage labor mobilization. Even though the Malaysian government in 1961 ratified the International Labor Organization's "Convention No. 98 on the Right to Organize," it has effectively struck down attempts of the Electrical Industry Workers' Union (EIWU), formed in 1974, to organize workers in the electrical and electronics industries (Wangel, 1988).

Until 1988, unions in Malaysia's microelectronics industry were effectively banned. When a union was formed at an RCA semiconductor assembly plant in 1984, the U.S.-based corporation (since taken over by General Electric) refused to recognize it and sold the company to the Harris Corporation (U.S.). Organizers who persisted in forming a company, not an industry-wide, union were harassed, fired, or reassigned with the implicit support of the Mahathir government. Over the years, the government has closely collaborated with the Malaysian-American Electronics Industry, an association with 16 electronics corporation members. When a union was registered by the government at Harris Solid State in 1990, the company was able to bypass the rules by simply not rehiring the unionized workers. In no case will the government permit either a national union or work stoppages, for fear, it claims, of deterring foreign investment in electronics, its biggest source of foreign investment, which, in recent years, has absorbed about 80% of all Japanese direct foreign investment. The government's strategy has been to create pro-UMNO labor formations, mainly white collar, and to permit, but not recognize or enforce, the rights of in-house unions as a way of controlling the blue-collar electronics workforce (Vatikiotis, 1992).

In the 1990s, pro-union attitudes have waned, according to Balakrishnan Nadison, an Electrical Industries Workers Union leader in Penang whom I interviewed in August 1995. This is remarkable given that Malaysia is experiencing a low level of unemployment (officially below 3% in mid-1995, below 4% in Penang) and a general labor shortage in electronics, a condition that normally should provide unions with greater bargaining leverage. However, the threat of using more foreign workers (at least 5,000 have already been recruited for Penang's electrical and electronics industries) and the overt support of the state gives corporations added latitude against labor mobilization. The government appears to have persuaded most workers that to be pro-union is tantamount to being antigovernment. The MTUC, the Malaysian affiliates of the International Metalworkers Federation (especially the EIWU and the Metal Industry Employees' Union) are considered by Mahathir to be "anti-national" (Jomo & Todd, 1994). Two companies—Harris Semiconductor (U.S.) and Hitachi Consumer Electronics (Japan)—refuse to recognize in-house unions within their premises, which they see as a potential obstruction to the goals of restructuring (Rasiah, 1995c).

Forms of Resistance

The main form of resistance to further encroachments on the prerogatives of electronics labor depends on unionization. Up to now, transnational corporations, in collusion with the state, have largely prevented unionization of the industry out of mutual concern that unions could undermine the objectives of export-oriented industrialization. When electronics has been organized, most effectively in the 1980s by the EIWU, it has been able to negotiate strong collective bargaining agreements with good wage increases and benefits (Wangel, 1988). The International Labor Organization has also been monitoring repression against Malaysian electronics workers and periodically has complained to the Malaysian government about abuses and illegal behavior against workers.

Women studies researchers and support groups in Malaysia have done the best work in publicizing the continuing and new forms of stress and the inequities experienced by female electronics shopfloor workers. Even when toxics and gases used in the semiconductor production process have been cut back or eliminated, there are still highly stressful job demands in working with assembly operations, especially in the area of consumer electronics. Apart from the ILO, women's organizations in Malaysia and internationally will be important in the monitoring of female labor in Penang's and other export processing zones. Gender issues bring another dimension to the struggle of the working class, and it is likely that as women increasingly come to regard their own labor as central to their or their families' survival, and not simply as "supplementary" income, they will make stronger demands for better wages and benefits. Working women in the U.S., a large percentage of whom are single parents, now represent the strongest force for unionization.

Malaysian women workers will also continue to force change in shopfloor conditions through actions that are culturally validated. Aihwa Ong (1987) noted that spiritualism did not end with the modern industrial transformation of Malaysia, and spirit visitations and "possession" among workers (similarly recorded by Grossman in her earlier Philippines study) recurs as a cultural means of protesting the dehumanizing conditions of work under capitalist social relations (see also Grossman, 1979). Resistance to the stresses and inequities of work in the electronics factories takes several other, sometime unanticipated, forms. Invoking Koranic injunctions to take prayer breaks during shift hours provides relief from the relentless pace of work. Other Koranic language alluding to the protection of women can also be referenced to demand that factory administrators and supervisors, especially foreigners, who are vulnerable to charges of ethnic and cultural insensitivity, accommodate the needs of the female workforce.

CONCLUSIONS: ELECTRONICS AND THE MALAYSIAN AND WORLD ECONOMY

Electronics production in Malaysia is part of the foreign, primarily U.S. and Japanese, enclave economy, is the most focussed area of foreign direct investment, and represents the main pillar of its export-oriented approach to development. Although electronics is associated with the advanced order of "high tech," Malaysia has received little technology transfer and skills training from this industrial sector in the 25 years since the launching of the country's first export processing zone. There are minimal linkages of electronics to other areas of the economy, with the exception of a few small-scale machine shops that service TNC firms. Such ancillary enterprises generate few jobs, about 16% of semiconductor firm employment in Penang as of 1985.

One reason for the constraining of ancillary employment is that the government's *bumiputra* policy pertains only to firms over a certain size. Given that most ancillary firms in Penang are Chinese-owned, the concern about ethnic tensions induces such entrepreneurs to retain small-scale operations. Furthermore, the fact that such firms conduct little R&D on their own and rely on the innovations of TNCs helps explain the lack of growth opportunities for local capital (Narayanan & Rasiah, 1992). Foreign capital investment is given priority over social or industrial capital formation, which, if not reversed, will most likely prove to be a critical flaw in Malaysia's development scheme.

At the same time, it is also true Malaysia has sustained high growth statistics almost continuously for three decades, and the short-term trend seems to indicate the economy's durability. The runaway yen to Malaysian currency (ringgit) exchange rate in recent years has encouraged a higher rate of Japanese imports from, and investment in, Malaysia. However, Malaysia's trade deficit keeps growing, reaching US$8.3 billion in 1994 alone, most of it with its leading trade partner, Japan. Growth in the Malaysian economy will eventually lead to removal of its GSP status, an action taken against the ANICs by the Bush administration. Loss of the ANICs' GSP status prompted U.S., Japanese, and European TNCs that export to the United States to relocate some of their investments to such surviving GSP havens as Malaysia and Thailand. Here, it was clearly state policy, not "comparative advantage," that drove investment and jobs to low wage production sites. Lower wage labor markets in other parts of Asia (China, Vietnam) is another threat to Malaysia's stability.

The new international division of labor plays workers against each other, while expanding the mobility of capital. When U.S. electronics corporations shut their doors to relocate to Southeast Asia, or elsewhere, as so many have done, thousands of workers lose their livelihood. Yet, as Storper and Walker (1989) have argued, "the actual practice of manufacturing at the periphery is a fundamental precondition for future advance [of industrialization and technological innovation]" (p. 117) in the core industrial states. The U.S. provides a better environment for labor struggles because of relative economic surplus, democratic traditions of the state, albeit limited, and the historic existence of a labor movement, albeit weak. In Malaysia, the state plays an important role in defending local class alliances with international capital, which requires it to suppress labor while reorganizing it along gender and ethnic lines to protect the interests of capital. The electronics industry is the door to Malaysia's participation in the international division of labor in communications equipment and services.

REFERENCES

Bangsberg, P.T. (1995, March 9). Malaysia becomes the rising star of US chipmaker's global operation. *Journal of Commerce and Commercial*, p. 3A.

Berberoglu, B. (1992). *The political economy of development: Development theory and the prospects for change in the third world*. Albany: State University of New York Press.

Brier, S. (Ed.). (1992). *Who built America?: Working people and the nation's economy, politics, culture, and society* (Vol. 2). New York: Pantheon Books.

Burawoy, M. (1985). *The politics of production*. London: Verso.

Cammack, P., Pool, D., & Tordoff, W. (1993). *Third world politics: A comparative introduction* (2nd ed.). Baltimore: Johns Hopkins University Press.

Chaponniere, R. (1984, November). The ASEAN integrated circuit: The electronics industry in ASEAN—Issues and perspectives. *ASEAN Economic Bulletin*, pp. 136-151.

Chee, H. L., & Subramaniam, M. (1994). Speaking out: Women workers talk about safety and health in electronics. In C. H. Leng (Ed.), *Behind the chip: Proceedings of the conference on safety and health in electronics* (pp. 81-102). Kuala Lumpur: Women's Development Collective Sdn. Bhd.

Cohen, R. (1987). *The new helots: Migrants in the international division of labor*. Aldershot, UK: Avebury.

Consumers Association of Penang. (1994, December 16). *Memorandum on the occupational and environment hazards in electronics industry*. Penang, Malaysia: Author.

Crouch, H. (1993). Malaysia: Neither authoritarian nor democratic. In K. Hewison, R. Robison, & G. Rodan (Eds.), *Southeast Asia in the 1990s: Authoritarianism, democracy & capitalism* (pp. 135-157). St. Leonards, New South Wales, Australia: Allen & Unwin.

Deyo, F. C. (Ed.). (1987). *The political economy of the new Asian industrialism*. Ithaca, NY: Cornell University Press.

Douglass, M. (1991). Transnational capital and the social construction of comparative advantage in Southeast Asia. *Southeast Asian Journal of Social Science, 19*(1 & 2), 14-43.

Fong, C. O. (1989). Wages and labour in the Malaysian electronics industry. *Labour and Society, 14* [Special Issue on High Tech and Labour], 81-102.

George, S., & Sabelli, F. (1994). *Faith and credit: The World Bank's secular empire.* Boulder, CO: Westview.

Ghazali, A. (1994, August 1). Electronics wage bill totals RM 1.65 billion. *Business Times,* p. 1.

Grossman, R. (1979). Women's place in the integrated circuit. *Southeast Asia Chronicle,* No. 66, and *Pacific Research, 9*(5 & 6), 2-17.

Haran, J. (1995, March 10). Penang draws electronics firms relocating from Singapore. *Business Times,* p. 1.

Hua, W. Y. (1983). *Class & communalism in Malaysia: Politics in a dependent capitalist state.* London: Zed.

INSAN. (1987). *Trade unionism for Malaysian workers.* Kuala Lumpur: Institute for Social Analysis.

Ismail, M. N. (1995, January). *Internationalization of capital and the Malaysian electronics industry.* Paper presented at the conference on Globalization: Local Challenges and Responses. Universiti Sains Malaysia, Penang, Malaysia.

Jomo, K. S. (1990). *Growth and structural change in the Malaysian economy.* New York: St. Martins Press.

Jomo, K. S., & Todd, P. (1994). *Trade unions and the state in peninsular Malaysia.* New York: Oxford University Press.

Kaur, L. (1995, January 12). Surge in TV, VCR exports to Japan. *Business Times.*

Keenan, F. (1995, November 23). Members only. *Far East Economic Review,* pp. 66, 70.

Kimori, M. (1989, September). Malaysia's workers: Jolting the electronics industry. *Multinational Monitor,* pp. 11-13.

Kreye, O., Heinrichs, J., & Frobel, F. (1988). *Multinational enterprises and employment.* Geneva, Switzerland: International Labour Organization.

Kua K. S. (1992). *Malaysian political realities.* Petaling Jaya, Malaysia: Oriengroup.

Levidow, L. (1991). Women who make chips. In *Science as culture* (Vol. 2, No. 10, pp. 103-124). London: Free Association Books.

Limqueco, P., McFarlane, B., & Odhnoff, J. (1989). *Labour and industry in ASEAN.* Manila: Journal of Contemporary Asia Press.

Lin, V. (1987). Women workers in the semiconductor industry in Singapore and Malaysia: A political economy of health. In M. Pinches & S. Lakha (Eds.), *Wage labour and social change: Proletariat in Asia and the Pacific* (pp. 219-261). Canberra: Monash University, Center of Southeast Asian Studies.

Markoff, J., (1996, January 9). World semiconductor market grew 48% in '95, report shows. *New York Times*, p. C13.

Mehmet, O. (1986). *Development in Malaysia: Poverty, wealth and trusteeship*. London: Croom Helm.

Momsen, J. H. (1991). *Women and development in the Third World*. New York: Routledge.

Narayanan, S., & Rasiah, R. (1992, October). Malaysian electronics: The changing prospects for employment and restructuring. *Development and Change, 23*(4), 75-99.

O'Brien, L. (1994). Some characteristics of work in the manufacturing sector. In H. Brookfield (Ed.), *Transformation with industrialization in peninsular Malaysia* (pp. 169-187). New York: Oxford University Press.

O'Donnell, G. A. (1973). *Modernization and bureaucratic authoritarianism in South American politics*. Berkeley: Institute of International Studies, University of California Press.

Ong, A. (1987). *Spirits of resistance and capitalist discipline: Factory women in Malaysia*. Albany: State University of New York Press.

Ong-Yeoh, D. (1995, March 8). Harris Corp. to expand operations in Malaysia. *Business Times*.

Penang Development Corporation. (1995, July). *Statistics: Penang, Malaysia*. Penang, Malaysia: Author.

Rasiah, R. (1989). Competition and restructuring in the semiconductor industry: Implications for technology transfer and its absorption in Penang. *Southeast Asian Journal of Social Science, 17*(2), 41-57.

Rasiah, R. (1994a). Facing the challenge: Trends and implications of the electronics industry in Malaysia. In C. H. Leng, (Ed.), *Behind the chip: Proceedings of the conference on safety and health in electronics* (pp. 11-29). Kuala Lumpur: Women's Development Collective Sdn. Bhd.

Rasiah, R, (1994b). Flexible production systems and local machine tool subcontracting: Case of Malaysia's electronics industry. *Cambridge Journal of Economics, 18*(1).

Rasiah, R. (1995a). *Foreign capital and industrialization in Malaysia*. New York: St. Martins Press.

Rasiah, R. (1995b, August). *Innovations and institutions: Technological learning in Malaysia's electronics industry*. Paper presented at the Innovation Networks: East Meets West conference. Sydney, Australia.

Rasiah, R. (1995c). Labour and industrialization in Malaysia. *Journal of Contemporary Asia, 25*(1), 73-92.

Salih, K., & Young, M. L. (1989). Changing conditions of labour in the semiconductor industry in Malaysia. *Labour and Society, 14* [Special Issue on High Tech and Labour], 59-80.

Scott, A.J. (1987). The semiconductor industry in South-East Asia: Organization, location and the international division of labour. *Regional Studies, 21*(2), 143-160.

Seward, N. (1988, October 6). UMNO's money machine. *Far Eastern Economic Review*, pp. 66-67.

Stockton, J, (1991). Poisoned factories & techno-fantasies in Malaysia. *Ampo, 23*(1), 62-67.

Storper, M., & Walker, R. (1989). *The capitalist imperative: Territory, technology and industrial growth*. Oxford: Basil Blackwell.

Tan, B. K. (1986). Women workers in the electronics industry. In H. A. Yun & R. Talib (Eds.), *Women and employment in Malaysia* (pp. 17-32). Kuala Lumpur: University of Malaya Press.

Taylor, M., & Ward, W. (1994). Industrial transformation since 1970: The context and the means. In H. Brookfield (Ed.), *Transformation with industrialization in Peninsular Malaysia* (pp. 95-121). New York: Oxford University Press.

Thornton, E. (1995, November 9). Logging onto Asia. *Far Eastern Economic Review*, pp. 68-72.

Vatikiotis, M. (1992, July 16). Credibility gap. *Far Eastern Economic Review*, p. 18.

Wangel, A. (1988). The ILO and protection of trade union rights: The electronics industry in Malaysia. In A. Wangel (Ed.), *Trade unions and the new industrialization of the third world* (pp. 287-305). Pittsburgh: University of Pittsburgh Press.

Warren, B. (1980). *Imperialism, pioneer of capitalism*. London: Verso.

7

Regional Subcontracting and Labor: Information/Communication Technology Production in Hong Kong and Shenzhen

Lai Si Tsui-Auch

The continuation of China's Open Door policy, the economic transformation of Eastern Europe, and the reopening of Vietnam's economy herald a genuinely globalized economy, reaching areas previously outside the orbit of transnational capital. Cities throughout the world are being wired by a new global communication/information order. Nation states are targeting microelectronics as one of the priorities in technological development and are seeking niches in rapidly changing technology markets.

In Hong Kong and the Shenzhen Special Economic Zone (SEZ), which I visited in Summer 1994, there is a growing popularity of information/communication technology (I/CT). In Hong Kong, mobile phones are used not only by businessmen and professionals but also by animated teenagers lost in conversation. One notices the proliferation of cellular telephones and the beeping of pagers. Sales of fax, telephone sets, pagers, and mobile phones are booming.

In Shenzhen, China, one sees emerging a modern metropolis, with scores of skyscrapers and congestion indistinguishable from neighboring, archcapitalist Hong Kong. Shenzhen is no longer the small border town and rural farmland it was in the early 1980s. People wear Hong Kong fashions now, listen to Hong Kong radio, and watch Hong Kong television programs via satellite dishes installed by local military units, which profiteered from these illegal sidelines (Overholt, 1993). The use of pagers is nearly as common in Shenzhen as in Hong Kong, and a mobile phone has become a status symbol for both local businessmen and cadre. Night life has come to resemble that of Hong Kong, with popular karaoke bars attracting locals and Hong Kong managers who reside in Shenzhen to supervise industrial operations.

Underlying these changing cultural pastimes, advances in I/CT have facilitated the coordination and control of manufacturing across the "borders" (Hong Kong to be reunited with China in 1997). The relocation of I/CT equipment manufacturing from Hong Kong to the SEZ and Pearl River Delta regions (see Figure 7.1) has been characterized by a technical and spatial division of labor, with highly capital- and skill-intensive processes concentrated in the First World and labor-intensive processes scattered around the world, especially in east and Southeast Asia.

This chapter, based on research data, interviews, and informal conversations with more than 40 manufacturers in Hong Kong and the Shenzhen SEZ, discusses the different labor policies and development strategies of the two cities in the global production of I/CT. Government departments in both locations classify I/CT as part of the electronics industry, and official statistics report on the electronics industry as a whole. Therefore, the macro data in this chapter refers to the electronics industry.

I argue that Hong Kong's regionalization of production, extended to southern China, has allowed small- to medium-sized firms and transnational corporations (TNCs) to continue to take advantage of cheap labor and to delay technological upgrading in the region. This has caused a loss of jobs among the Hong Kong workers and a strong reliance on a migrant labor force from inland China. Migrant workers toil long hours at low wages and with minimum welfare protection. The adverse conditions faced by the workers reflect the cost-saving strategies of manufacturers and the compromise with foreign capital by the Chinese authorities in its march toward global integration.

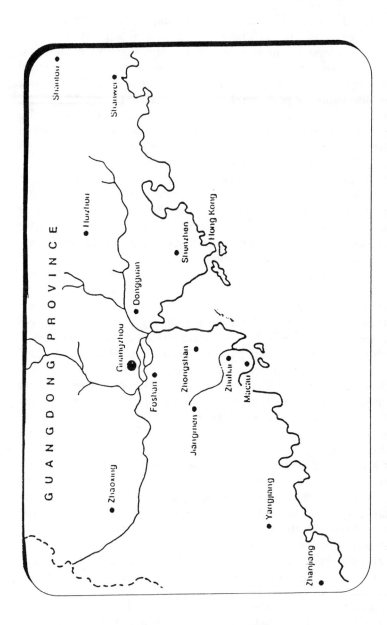

Figure 7.1. Location of Hong Kong, Shenzhen, and the Pearl River Delta Regions.
Source: In Hong Kong Investment in Pearl River Delta published by Hong Kong Federation of Industries (1992).

REVISITING THE DEBATE ON LABOR AND GLOBAL PRODUCTION

Both dependency (Frank, 1967, 1981) and world system theorists (Wallerstein, 1974, 1979) recognize the existence of a geographically differentiated division of labor within the capitalist world order—between the core and the periphery or the core, semiperiphery, and periphery. To Wallerstein, the semiperiphery helps sustain the continual dominance of the core. The existence of semiperipheral (newly industrializing) countries helps to demobilize what would otherwise become an opposition of peripheral countries against the world system, with the semiperiphery's stake constituting a shared role in the exploitation of peripheral countries. In addition, they present themselves to peripheral countries as examples of the possibility of upward mobility, thereby creating legitimacy for transnational capitalism.

Wallerstein conceptualizes all countries as part of a single world system but, paradoxically, takes the nation state as his unit of analysis. However, the formation of regional economic blocs and the global integration of export-oriented cities would appear to offer a more appropriate focus than that of national economies, requiring, therefore, alternative conceptions and perspectives.

The uneven development of world cities resulting from the expansion of transnational capital has been substantively discussed by urban theorists. Hymer (1972) posits a spatially uneven development among cities resulting from TNCs' global strategy of production. He conceptualizes the world system as a TNC-dominated spatial division of labor, organizational hierarchy, and structure of domination/subordination. The geographical dispersion of production mirrors the organizational evolution of TNCs toward an increasingly complex and efficient multiproduct, multidivisional structure. Cities emerge as the points of organization for TNC activities, and different cities play different roles. The top-level activities are located in capital cities, close to financial markets and governments; middle-level administrative activities are located in large cities to tap white-collar workers and communication systems; and extractive, manufacturing, and sales activities are located throughout the world depending on the need for labor, markets, and raw materials. Hymer's effort to explicate the relationship between cities and transnationals' activities is further elaborated by Friedmann and Wolff (1982).[1]

Whereas Hymer and Friedmann and Wolff elaborate the global economy in terms of spatial relationships, Fröbel, Heinrichs, and Kreye

---

[1]Friedmann and Wolff (1982) hypothesize the spatial articulation of the emerging world system of production and markets through a global network of cities called "world cities."

(1980) discuss it in terms of production relations. Fröbel et al. see the international division of labor since the 1960s as increasingly based on "process specialization" between the developed and developing countries. That is, TNCs have transferred labor-intensive processes to low-cost Third World countries and specialize in the capital- and technology-intensive processes in their home countries.

The emergence of this international division of labor has been facilitated by preconditions of the development of a worldwide reservoir of potential labor power (often a politically repressed and subjugated labor force), the decomposition of complex production processes into separate and simple units of production that can be carried out by unskilled labor with little training, and the improvement in transportation and communication technologies, which free industrial production from the constraints of geographical location and allow TNCs to take advantage of wage differentials in various locations worldwide. Consequently, a growing proportion of world trade is composed of flows of commodities between plants of the same company or between contractors and subcontractors in various parts of the world. This new international division of labor (NIDL) is characterized by an ongoing process of increasing subdivision and geographical separation of manufacturing processes, with different labor requirements and means of labor control. Thus, Third World countries are incorporated as segments of centrally coordinated "global factories" headquartered in the core capitalist states. The relocation of assembly processes has led to both structural unemployment in the First World and labor exploitation in the Third World.

The NIDL thesis has been enriched by feminist and ethnic perspectives (Fuentes & Ehrenreich, 1984; Hollander, 1977; Sassen-Koob, 1980). The electronics/semiconductor industry has been a prototypical example of the thesis (Aldana, 1989; Eisold, 1982; Elson, 1981; Grossman, 1979; Katz & Kemnitzer, 1983; Sivanandan, 1979). However, these perspectives have been strongly criticized for their analytical inadequacies with respect to technological changes in labor processes and perceptions of "exploitation."

Researchers argue that advances in microelectronics and automation technology render the cost-cutting strategy in offshore production less significant. The new technologies induce some degree of "deinternationalization." The onshore move of manufacturing in semiconductor, consumer electronics, and computer industries has been noted[2] (Rada, 1980; Sayer, 1986).

---

[2]U.S. companies such as Motorola, Intel, Fairchild, Applied Microcircuits, Apple, IBM, and Prism Co. established onshore automated assembly plants. Some Japanese companies withdrew from South Korea and Taiwan and established their own onshore assembly plants (Sayer, 1986).

With technological and organizational innovations, "flexible specialization" (Piore & Sabel, 1984) tends to replace "mass production." Companies extract greater added value through production for specialized markets. Flexible specialization combines the advantages of craft flexibility with the most advanced information processing and microelectronics technology. Companies organized into flexible production systems and located near specialized markets are able to respond quickly to market fluctuations. Because specialized markets concentrate in First World countries, the shift to low-wage sites will not occur on the scale once expected.

However, counterevidence to the deinternationalization argument also has been presented. Automation has not driven TNCs to relocate onshore to any great extent. First, as automated machines are expensive, they are cost effective only in 24-hour use cycles, based on multishift, seven-day workweek scheduling. Countries with minimum labor and environmental regulation are ideal testing grounds for introducing automation. Second, a rapid retreat to a less internationalized mode of electronics manufacturing involves substantial costs in closing down existing plants and reshuffling supply and market networks. Third, Asian newly industrialized countries (ANICs) have a sufficient supply of low-cost engineers, technicians, and skilled workers for automated operations. Automation reduces employment of unskilled labor but increases the employment of skilled labor (Ernst & O'Connor, 1992).

It is argued that manual assembly is still cost effective for less sophisticated small-batch products (Sanderson, Williams, Ballenger, and Berry, 1989). Automation has not been widespread in offshore assembly of television sets because output defects can be corrected at relatively low costs (UNCTC, 1987). Since the yen shock of 1985-1986, some Japanese companies have lost confidence in their ability to compete solely on the basis of automated technologies. It is evident that many Japanese corporations are relocating offshore (*Far Eastern Economic Review*, 1995, various issues; Mody & Wheeler, 1990; Yamada, 1990).

Aspiring ANICs such as Malaysia, the Philippines, Thailand, Indonesia, and China, with cheap labor supplies, will continue to carry out low value-added assembly processes in accordance with a new international division of labor theory. Despite their rising labor costs, the ANICs (South Korea, Taiwan, Singapore, and Hong Kong) have an edge in component production, engineering skills, and market potential and are likely to attract foreign automated operations in electronics manufacturing and simple design processes. They have improved considerably their technological capacity over the years, which is contrary to the hypotheses of NIDL theorists. These are the first signs of a new hierarchy in the global production system (Sanderson et al., 1989; Ernst & O'Connor, 1992). The ANICs have taken over the production of lower end products both in original equipment manufacturing (OEM) and own brandname manufacturing (OBM), whereas aspiring ANICs remain subsidiary bases of TNCs. Japan, the ANICs, and aspiring ANICs have been moving toward a multilayered structure based on both vertical and horizontal specialization (Yamada, 1990).

It is simplistic to generalize labor exploitation without taking into consideration the labor processes of a particular industry in a time-space continuum. Heyzer's (1986) research demonstrates differences in the structures and labor processes in electronics and textile industries in Singapore. Both employ mainly female labor for assembly. However, the semiconductor industry is controlled by TNCs and tends to use more automated and clean room technologies. They employ women of higher educational levels and mostly local workers. The textile industry is operated mainly by local capital working as subcontractors for TNCs. In sweatshops, which coexist with mass assembly factories, homework is commonplace. They usually offer lower wages and employ mostly migrant workers (from Malaysia), housewives, and children.

Researchers have often defined the concept of labor exploitation in objective terms such as poor levels of wages, conditions of work and housing, and excessive social controls in the workplace. However, workers not finding alternative livelihoods may be willing to accept factory work and may not perceive such employment as exploitative. Heyzer (1986) finds that "the provision of wage employment away from home and the patriarchal family system is generally viewed by the women themselves as 'liberating'" (p. 110). In his study of the Shenzhen SEZ, Sklair (1991) argues that one should not be ethnocentric and moralistic about this issue. A revisiting of the debate on labor exploitation suggests further research on the emerging regional division of labor and the impacts on labor of technological change, production structure, and the composition of capital.

THE CHINESE MODE OF ELECTRONICS PRODUCTION

From its inception as a British colony until the mid-20th century, Hong Kong was incorporated into the world economy as an East Asian entrepot. The colony, early integrated into global manufacturing, was well known for its cheap labor in the 1950s and 1960s. Its manufacturing base has gradually given way to a service economy since the mid-1970s, and the city has been developed into a global financial center. With the recent open economic policy of China, Hong Kong has become the largest investment, export, funding, financial, and information center for the People's Republic.

Hong Kong is densely populated, with almost 6.4 million people (98% of whom are Chinese) crowded into 1,000 square kilometers of mountainous land on Hong Kong island, Kowloon, and the New Territories. Hong Kong is highly deindustrialized, with industry accounting for 16.9% of GDP, service for 83%, and agriculture and fishing for 0.24% in 1994 (Hong Kong, 1996). Hong Kong attained an average annual economic growth rate of 5.5% from 1991-1995 (UNESCAP, 1997). Its per capita income of about US$23,000 in 1995 placed it second in Asia just behind Japan (Hong Kong Yearbook, 1996, p. 432).

Shenzhen is located at the southern end of Guangdong Province, across the border from Hong Kong. An area of 327.5 square kilometers (17% of the physical size of the municipality) was enclosed to be a SEZ in 1980. The SEZ is "special" because it is the first area (together with three other SEZs in Zhuhai, Xiamen, and Shantou) in China in which free market mechanisms and foreign participation are promoted. Shenzhen has been a cheap labor haven for Hong Kong industrialists throughout the 1980s; thus, Hong Kong manufacturers constitute the largest investing group, followed by Taiwanese, Japanese, and Western manufacturers.

When Shenzhen SEZ was established in 1980, the whole municipality had 333,000 residents. By 1993, the population increased to 2.9 million, including 2 million temporary residents who account for 70% of the total. The SEZ has 1.2 million residents, with a temporary population of about 668,000, accounting for 56% of the total population for the SEZ.

Most of Shenzhen's production and trade are concentrated in the SEZ. In 1993, the GDP of Shenzhen amounted to US$4.7 billion and the previous year per-capita income was about US$1,730. The GDP of the SEZ alone amounted to US$3.6 billion, with per-capita income at US$3,075, about one sixth that of Hong Kong (see Table 7.1).

HONG KONG AND SHENZHEN IN THE
GLOBAL PRODUCTION OF I/CT

Global competition in electronics production has accelerated technological changes and industrial restructuring among global electronics corporations. In the 1970s, the single most important technological change in the electronics industry was the advance in microelectronics, which made possible the automation of assembly and quality control processes. Consequently, electronics production in First

Table 7.1. Basic Data of Hong Kong, Shenzhen Municipality, and SEZ.

	Hong Kong	Shenzhen Municip.	SEZ
Area (sq. km)	1,000	2,020	327.5
Total Population (Mil.)	5.9	2.9	1.2
temporary	N/A	2.0	0.7
permanent	N/A	0.9	0.5
Contribution to GDP (%)			
primary sector	6	3	1
secondary sector	18	56	55
tertiary sector	76	41	44
GDP (US$ billion)	109	4.7	3.6
GDP/Capita (US$)	18,907	1,730	3,075
GDP increase in real terms (%)	5.5	> 30	
Inflation rate (%)	8.5	35	

Sources: Data compiled from *Hong Kong Yearbook* (1994, pp. 50-53); *Shenzhen Yellow Pages Commercial/Industrial Directory* (p. 613); *Directory of Investment in Guangdong* (1993, p, 11).

World countries has become much less labor-intensive than before. Higher quality products are produced with the use of computer-aided design, computer-aided manufacturing systems, and industrial robots. The center of competition has thereby shifted beyond the sphere of production to R&D and to the coordination of increasingly complex network transactions with component suppliers, customers, and external technological sources (Ernst & O'Connor, 1992).

Microelectronic applications reduce the significance of labor costs and increase the capital intensity and the advantage of proximity to niche markets concentrating in the First World. In the 1980s, there were some signs of deinternationalization of electronics production. Growth of foreign direct investments in electronics (especially in semiconductors) in Asia was rather slow.

Nevertheless, the diffusion of electronics production to Asia has continued. TNCs retain operations in the ANICs and aspiring ANICS to take advantage of the abundant supplies of low-cost skilled labor and the established regional sourcing networks. Those Asian manufacturing centers with minimum labor regulations are seen as ideal sites to practice automated assembly on a 24-hour cycle, multishift, seven-day week schedule.

The use of automation is uneven among different processes, products, and electronics-producing countries and is more prevalent in semiconductor than consumer electronics manufacturing. In semiconductor production, automation prevails more in wafer fabrication and testing processes than assembly processes. Automated production processes also are seen more in the ANICs than the aspiring ANICs. Abundant supplies of low-wage labor in Third World countries has slowed the introduction of automation and encouraged less expensive semiautomated manufacturing processes in offshore TNC subsidiaries.

While concentrating resources in developing new products and markets, TNCs have also increasingly subcontracted to NIC companies to produce mature products on an OEM basis. They are focusing on higher value-added products (integrated circuits, computers, and multiple-function videotape recorders), whereas ANICs are specializing in lower value-added goods (TVs, audio equipment, computer peripherals, watches, and calculators). This is called product specialization (an unequal horizontal division of labor). Concurrently, the ANICs are developing process specialization, or a vertical division of labor with the Southeast Asian region and China, in order to reduce production costs.

The integration of Hong Kong into the production of I/CT goes back to the late 1950s, earlier than South Korea, Taiwan, and Singapore. Hong Kong became the second Asian site (after Japan) to attract U.S. offshore investments and the first location to attract Japanese electronics investments. Intensifying competition with U.S. electronics producers in the late 1950s and growing wage rates in Japan helped induce Sony to partially internationalize its radio assembly. The result was that Hong Kong's first electronics firm, Champagne Engineering Company, assembled in 1959 more than 4,000 radios a month for Sony under a subcontract arrangement (Henderson, 1989). By 1960, Champagne and two other companies had started to manufacture their own radios at a cost even lower than that of the Japanese (resulting from lower labor and overhead costs). Consequently, Hong Kong radio exports to the United States effectively undercut the Japanese at the lower end of the market during 1960-1961.[3]

---

[3]By 1961, 12 companies manufactured radios in Hong Kong, of which two were joint ventures with U.S. companies (Ng, 1992). The successful competition with Japanese producers led to a Japanese government ban on the export of transistors to Hong Kong in 1962. They were, however, substituted by imports from the U.S. and Britain. Meanwhile, Hong Kong's radio output increased by one third that year. The Japanese government realized that its ban was ineffective and lifted it the same year. Furthermore, the Hong Kong companies moved quickly to manufacture condensers, capacitors and transistors for radio production in 1962. The radio manufacturing industry flourished, and the number of companies increased in the following years (Hong Kong Government Industry Department, 1993).

Electronics production was diversified to FM radios (1964), television tuners (1965), and so on. Companies next moved on to produce television sets and tape recorders, as well as parts and subassemblies for these products. The manufacture of computers, starting in the early 1980s, was not technology-intensive. Hong Kong assembled mostly microcomputers for the very low end of this market.

In the 1960s, a number of semiconductor manufacturers established assembly plants in Hong Kong. Fairchild Semiconductor, the "mother" of Silicon Valley, moved assembly operations to Hong Kong in the face of substantial price competition in transistor markets. Motorola, National Semiconductor, and Sprague followed. By the late 1960s, Hong Kong had become the principal Asian assembler of semiconductors for the U.S. market. Japanese companies such as Hitachi, Oki, and a number of others also moved assembly plants to Hong Kong, whereas several local companies emerged to assemble chips for TNCs (Henderson, 1989).

Most Hong Kong manufacturers work as OEM suppliers for TNCs, meaning they produce according to specifications given by the contractors and do not need to promote themselves in foreign markets or carry on R&D. Foreign contractors send personnel from headquarters or employ agents to inspect OEM manufacturers' production technology, factory layout, sales records, and so on before placing an order. After production, the inspecting agents examine the products to be shipped to contractors. Recent high commission rates for agents have cut the profit margin of low-priced product manufacturers. Japanese manufacturers do not place OEM orders with Hong Kong manufacturers unless there is a 30% to 40% saving in production costs as compared to domestic manufacturing. This amount of cost saving is required to cover offshore procurement risks and to allow for additional transportation, insurance, quality inspection, and warehousing expenses (Hong Kong Trade Development Council [HKTDC], 1991).

The electronics industry has been facing structural problems, including rising wage and land costs, low technology standards, weak supporting parts and components segments, severe local and overseas competition, and an overreliance on the U.S. market and on Japanese supply of critical components. Small- to medium-sized manufacturers spend about 60% of the production cost on components such as ICs, imported mostly from Japan and, to a lesser extent, South Korea and Taiwan. Despite its early entry into electronics production, compared to other NICs, Hong Kong still depends on Taiwan and South Korea for supplies of critical components and parts (Ho, 1992).

Nearly 80% of Hong Kong electronics manufacturers maintain their competitiveness by moving assembly processes across the Hong Kong border to Shenzhen and the Pearl River Delta region (see Figure

7.2). Some TNCs (long based in Hong Kong) such as Sanyo, Philips, Commodore, and AST have adopted the same strategy. They also take advantage of preferential entry agreements with the U.S. under the Generalized System of Preferences (whereas Hong Kong lost GSP status after January 1988).

In 1978, there was only one electronics assembly plant, employing 100 workers, in Shenzhen, and it produced only simple radios, amplifiers, and transformers. By the end of 1983, electronics became Shenzhen SEZ's largest industry, as Hong Kong-based manufacturers moved their production across the border. In 1985, the electronics industry constituted 48.7% of Shenzhen SEZ's total industrial output (Chen, 1987). Assembly and processing branches of electronics grew at the expense of other industries and against the Chinese government's desire for industrial diversification.

By 1992, electronics was Shenzhen's largest industry, accounting for about 31.7% (34.8% in the SEZ) of the gross industrial output of the municipality and 13% of China's total electronics industry output value (*Shenzhen Yellow Pages Commercial/Industrial Directory*, 1993, p. 690). The industry has been the largest employer in the manufacturing sector, employing 123,171 (83,124 in the SEZ alone), representing 22% of the total employment in the whole municipality. Production in Shenzhen SEZ and the municipality is much more labor-intensive than that of Hong Kong; in

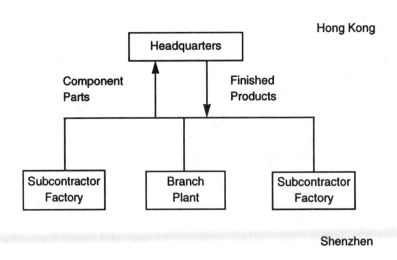

Figure 7.2. Subcontracting

1992, the average number of workers per establishment in Hong Kong was 42,227 in the SEZ, and 345 in the municipality (see Table 7.2).

INDUSTRIAL SUBCONTRACTING AND LABOR

The relocation of labor-intensive production processes to the SEZ mirrors a spatial extension of industrial subcontracting, a practice whereby a company offers subcontracting requests to another independent company to produce or process a component or part according to the specifications it provides. Subcontracting differs from the mere purchase of ready-made parts and components from suppliers in that there is an actual contract between the two participating companies setting out specifications for the order (Friedman, 1977).

The Hong Kong management of capital-rich companies tend to favor operations that are small in terms of capital outlay, although the workforce may be large because of the labor-intensive nature of production. With small operations, manufacturers find it faster to secure licensing approval and easier to get around the Chinese bureaucracy in tax payment and labor recruitment. This mode of operation is termed *guerrilla capitalism* (Smart & Smart, 1991).

Table 7.2. Basic Data of Electronics Industry in Hong Kong and Shenzhen.

	Hong Kong (1993)	Shenzhen Municipality	Shenzhen SEZ (1992)
No. of companies	1446	357	300
No. of workers	53,591	123,171	83,124
Average No. of workers/company	42	345	227
Contribution to manufacturing employment (%)	8.9	22	N.A.
Contribution to total industrial value (%)	N.A.	31.7	34.8
Export Value (million US$)	7,350	995	840
Export as total domestic export (%)	25.7	N/A	N/A

Sources: Data compiled from *Hong Kong Annual Yearbook* (1994, pp. 464-466); *Shenzhen Yellow Pages Commercial Industrial Directory* (1993, p. 670).

To support production capacity, companies resort to extensive use of subcontracting[4] (see Figure 7.2). The companies, as contractors, maintain a supervisory role over subcontractors and supply them components, training, and production equipment. The contractors' goal is to keep a stable and long-term relationship with subcontractors, who, in turn, rely on contractors' orders.

The roles of Hong Kong and Shenzhen SEZ in the production of I/CT have changed over the years. In the early 1980s, companies operating in Hong Kong supplied materials to SEZ factories only for processing or assembling. The Shenzhen subcontractors earned processing fees when they transported the semifinished products to Hong Kong manufacturers, who did the final assembly, testing, and packaging of the finished products.

In recent years, SEZ authorities have pressured foreign companies to upgrade their investment from processing and assembling arrangements to joint ventures with Chinese partners or wholly foreign-owned subsidiaries when the investors applied for renewal of investment contracts. Many moved north to the Bao'an District or even further outside Shenzhen to other Pearl River Delta cities. Those who remained in the SEZ had to comply with the new demands.

Hong Kong companies either set up subsidiaries in the SEZ or bought up existing SEZ factories. Most companies transferred all production processes, including assembly, testing, and packaging, to the SEZ factories, while some retained a small workforce in Hong Kong to do pilot runs and packaging. The branch plant maintains a management capacity for recruitment, worker management, handling customs procedures, training, and technical support. However, these functions are vertically integrated by the Hong Kong headquarters, in which the control of production and management is centralized (Figure 7.3).

As product manufacturers relocated their production to the SEZ, component suppliers also moved across the border to ensure timely delivery. Electronics manufacturers can simply place orders at the suppliers' Hong Kong offices, and their branch plants in the SEZ can get the delivery directly from the branch plants of the suppliers in the same locality (see Figure 7.3). The products for export are transported from the branch plants directly to the container ports in Hong Kong, without going through the Hong Kong headquarters. The production of the end-products and components have been largely relocated out of Hong Kong.

---

[4]Similar to what has been characterized as "capacity subcontracting" (Vennin & de Banville, 1975), "concurrent subcontracting" (Scott, 1983), or "cyclical subcontracting" (Watanabe, 1971), the Hong Kong-based contractor and the Chinese subcontractor engage in similar work and are mutually competitive by nature.

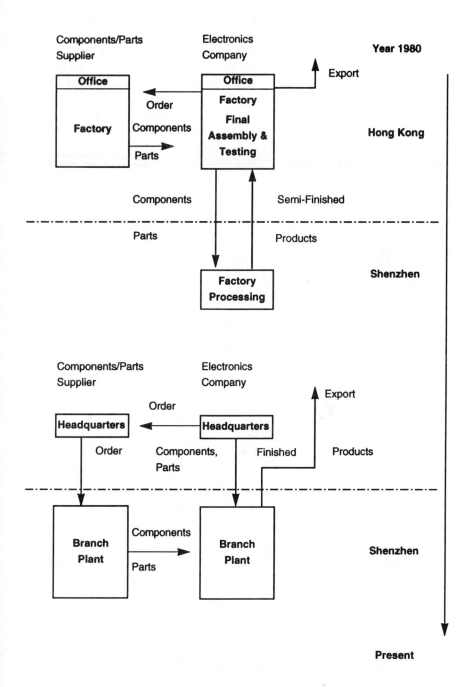

Figure 7.3. Changing intra-firm and inter-firm division of labor

THE DISPLACEMENT OF HONG KONG WORKERS

Decades of labor struggle, together with an increasing economic prosperity, have led to more protection and welfare for workers.[5] Hong Kong's industrial wage, the second highest in Asia after Japan, has increased from about US$38 in the 1970s to more than US$500/month (four to seven times higher than the wage rate of the SEZ) with associated welfare benefits. Some manufacturers admitted that they employ Chinese immigrants for assembly work in Hong Kong. The massive relocation of production processes has led to a significant reduction of employment. Out of 57 TNC electronics subsidiaries listed in the *Directory of Hong Kong Electronics Industry* (HKEA, 1994), 26 have regional headquarters or offices, 28 carry out sales, but only 13 maintain production in Hong Kong. The hypothesis that TNCs will upgrade manufacturing in the NICs, while relocating labor-intensive processes to less-developed economies, is not supported in the case of Hong Kong.

The impacts of restructuring in various branches of the industry are not uniform; workforce reduction is larger in branches in which the production is more labor-intensive. In a survey of the electronics producers in Hong Kong, I found that radio, television, and

---

[5]Women and young workers (from 15 to 17 years old) are by law protected from not working more than 8 hours a day and 44 to 48 hours a week. Those under the age of 18 are prohibited from working overtime. The overtime work for women in industrial employment is restricted to 2 hours a day and 200 hours a year. Women and young workers must be given one day off per week.

Most factories are closed on Sunday, and some adopt a five or five-and-a-half-day week scheme (44 to 48 hours per week) without any reduction in wages. Employees are entitled to 11 statutory holidays a year if they work continuously for an employer for three months preceding a holiday. All workers are entitled to seven days' annual leave with pay after working for a year under a continuous contract with the same employer. Sickness allowance is reduced to two thirds of a worker's pay for a sick leave period of more than three consecutive days.

A female worker who has served the same employer for 26 weeks is entitled to 10 weeks' maternity leave. If she has 40 weeks' service and no more than two surviving children, she is entitled to pay during her maternity leave of two thirds of her normal wages. When a pregnant worker entitled to maternity leave has completed 12 weeks' service with the same employer and gives notice of her intention to take maternity leave, the employer is prohibited from terminating her contract of employment until the date on which she resumes work after the maternity leave (Chen, 1991). A violation of such regulations induces severe punishment.

The employers use various methods to replace experienced, higher salary workers with newer workers. Companies remove the long-serving workers to avoid paying salary increases and end-of-the-year bonuses.

communication equipment manufacturers dismissed all workers and closed their factories in Hong Kong, while IC manufacturers retained a workforce for major production processes. The employment in the parts and components branch fell from 25,702 in 1991 to 21,792 in 1993, an average annual reduction of 7.9%. In the same period, radio, television, and communication equipment employment fell from 10,129 to 5,654, an average annual reduction of 24.8% (*Hong Kong Yearbook*, 1994).

The number of firms reached a peak of 2,009 in 1989 and decreased steadily to 1,446 in 1992 (Hong Kong Government Industry Department [HKGID], 1993; see Table 7.3). The number of employees reached a peak of 109,677 in 1989 and dropped at an average rate of 13.5% to 53,591 in 1993 (HKGID, 1993). Electronics factories in the mid-1960s to mid-1970s were large, usually employing more than 100 persons, but the relocation to China and increased automation led to a significant drop in average factory size. In 1992, the average number of workers per establishment was 42 persons (HKGID, 1993).

Table 7.3. Firms and Employees in Electronics Industry in Hong Kong.

Year	No. of Persons % of Manufacturing Industries	No. of Persons % of Manufacturing Workforce	Avg. No. . Persons/Co. Electronics Industry	Avg. No. Persons/Co. All Manu-Industries
1965	35 (0.4)	5,013 (1.5)	143	39
1975	490 (1.6)	53,833 (7.9)	110	22
1985	1,304 (2.7)	86,115 (10.1)	66	18
1988	1,939 (3.8)	109,677 (13)	57	17
1989	2,009 (4)	99,455 (12.4)	50	16
1992	1,446 (3.4)	60,653 (10.6)	42	14
1993	N/A	53,591 (11)	N/A	N/A

Source: Data compiled from Hong Kong Labor Department, *Employment Statistics*, 1993; Government of Hong Kong, *Hong Kong 1994*.

CHINESE WORKERS IN THE SEZ

Production relocated to the SEZ is confined to assembling and testing. Assembling is largely labor-intensive, in which workers fit hundreds of tiny parts and components, using "tools" (such as stone ovens and wire-cutting implements) rather than "machines." The production of color televisions involves automated assembly lines and equipment of greater precision than that of radios. Testing involves more sophisticated and semi-automated equipment. Nevertheless, the level of automation is generally low. Manufacturers explain that automation is meant for standardized mass production and is not cost-effective for small-scale production. The orders are usually small, and each contractor has slightly different specifications in function and appearance of the same product.

TNCs do not necessarily employ more sophisticated equipment. They aim mainly at taking advantage of cheap labor to produce for exports but not at penetrating the Chinese market. They, therefore, are not under any direct pressure to bring in advanced technology. Sanyo's operations were largely labor-intensive until the mid-1980s. However, their more recent installation of relatively advanced technology in Huaqiang Sanyo, a joint venture with a state enterprise, and other recently built subsidiaries of Sony indicate interest in selling a higher proportion of product output in the Chinese market.[6]

The labor-intensive nature of most investment has created a high demand for low-wage workers. Shenzhen was a small border town before 1980 with a sparse population. Over the years, high economic growth and increasing production investment and trade led to an expansion of the service sector, enticing local young women to work in hotels, restaurants, and shopping centers, instead of factories.

Thus, foreign investors increasingly employed migrant workers from other localities within and outside Guangdong Province. Among the electronics manufacturers I interviewed, only three reported that they rely on a labor service company and their Chinese partners to recruit workers, most of whom are from within the province. All other manufacturers stated that more than 70% of their workers are recruited from outside Guangdong.

The employment policy of Guangdong Province stipulates that workers from within the province have priority over those from outside, as do workers from mountainous areas over those from the cities. The policy is meant to lend support to the province, especially poor regions.

[6]Sanyo is permitted to sell 100% of its VTR production to the Chinese market, 50% of the color televisions and 25% of tape recorder production. Sanyo has built seven subsidiaries in Shenzhen and plans to keep its operation if social and political stability is maintained.

Migrant workers constitute 90% of the manual workers of Shenzhen municipality; 60% of them are 15 to 24 years old, and 73% are female (*Hong Kong Economic Journal*, January 1994, p. 40). Employers seek workers mostly from Qinghai, Guangxi, Hunan, Sichuan, Jiangxi, Zhejiang, and other provinces further north, as they demand lower salary and tend to be more compliant. In order to look for jobs in coastal areas, some of these workers from rural areas must pay a year's salary to local cadres for permission to travel. Employment of workers from outside Guangdong has to be reported to and approved by the labor bureau of the province.[7] In practice, local governments normally approve of the labor brought in by foreign investors after the fact.

To avoid the expense of getting "contracted workers,"[8] companies prefer to use migrant workers who do not ask for contracts. Manufacturers defend themselves by arguing that workers do not observe the contracts anyway; many simply quit when they return to their home villages for Chinese New Year vacation. Chinese managers and staff are employed at an average salary of RMB2,000 (US$230), engineers at RMB1,500 (US$170), and direct operators only RMB600 (US$70). Managers and engineers are given contracts, but operators are not. Investors get around the regulations by using personal connections and paying for protection.

The "temporary urbanization" as a result of the inflow of migrant workers is an added burden to the already inadequate infrastructure. To cope with the inflow of migrant labor, Chinese authorities demand that manufacturers pay the costs of social infrastructure. Companies usually offer canteen facilities and dormitories with two square meters of space for each migrant worker. Beyond that, companies are asked to pay a monthly management fee of about US$1 per worker and an annual social infrastructure charge of

---

[7]According to a Shenzhen Municipal government study of more than 5,000 companies, more than 600 factories have not reported to the government about their recruits. More than 70,000 workers are not registered, and child labor are frequently employed (Yu, 1996). Another widely reported form of corruption is the paying of bribes to local village and town authorities by workers seeking permits to leave their homes to work in Shenzhen.

[8]Foreign investors are required to pay for a contracted worker a percentage (15% to 23%) of the wage into a labor insurance fund, which offers retirement benefits for workers; 1% of the wage as unemployment insurance, 4% to 8% as health insurance, 0.8% to 2.5% as injury insurance. A *Workers' Daily* story reported a survey of Shenzhen's Futian district that showed that more than half of 465 companies have not paid accident insurance. In a neighboring district, two thirds have not paid unemployment insurance fees. Some factories also routinely neglect to have worker's contracts validated by local authorities, leaving sacked workers no resource. (cited in Goldstein & Huus, 1994).

about US$40. There are other charges such as fees for "keeping sanitation" and "growing trees," but such extraneous costs are still lower than those found in Southeast Asian countries.

Fees are frequently imposed without prior notice, making it difficult to carry out forward planning, which induces manufacturers to skirt laws and regulations and to force lower wages in anticipation of new cost requirements. In fact, one of the reasons why companies maintain small operations is to better cover up illegal practices. A smaller workforce is easier to manage, and in this situation workers cannot readily organize collective actions against management.

To support production capacity, manufacturers encourage employees to work overtime for wages paid in Hong Kong dollars and/or subcontract a part of production to smaller companies. Some subcontractors are overwhelmingly dependent on the orders of a single contractor. To guard against market fluctuations, manufacturers usually employ a workforce smaller than required and work them overtime during busy seasons. In my survey, all 40 manufacturers interviewed except two (who observe the 5 and 1/2 day/week rule)[9] work on a six- to seven-day week schedule. Workers are asked to work overtime for two to four hours per day. Many manufacturers stay at their Hong Kong offices Monday to Thursday and visit their factories Friday to Sunday.

Workers are required to undertake a few days of on-the-job training related to their tasks. Whereas most TNCs set up training departments, most Hong Kong companies do not have training departments and do not keep records of workers. Systematic training is seen as unnecessary given the low skill requirements of production tasks and the high mobility of the workforce. Most workers see no long-term prospects in their jobs.

The relationship between workers and foreign management is far from harmonious. Andors (1988) and Leung (1988) documented a strike organized by workers at a Sanyo plant in 1987 in Shekou Industrial Zone, Shenzhen. Japanese management complained that Chinese workers were interested more in wages than in skill development. One manager criticized the lack of female cooperation

---

[9]Clauses protecting workers' welfare in Guangdong Province are largely not observed by the employers except for Huaqiang-Sanyo. The clauses (*The Contemporary*, 1994) state: (a) a worker who is employed for more than 30 days is entitled to receive a contract from the employers; (b) a worker is by law protected from not working no more than 5 1/2 days per week and 8 hours per day; and (c) workers are entitled to the right to not work overtime. The overtime work is restricted to 48 hours per month. An employer has to get the consent of workers when he/she requires them to work more than the allotted time limit and to report to the local labor service company.

during tea breaks and expressed the need to teach Chinese female workers "appropriate manners" (meaning Japanese-style tea-serving practices). Workers complained that Sanyo pushed them too hard on quality control and efficiency standards and maintained an intruding surveillance system within the factory.

Manufacturers often complain that Chinese workers are "poorly educated," "greedy," and "unproductive." They attribute the inefficiency of Chinese workers to their work "mentality" and low education. One telephone set manufacturer complained:

> They think that the boss is exploiting them, and they feel like prisoners. Discipline is hard to enforce. For example, I have tried to stop workers from spitting in the workplace. In the beginning, I imposed a penalty of 10 Yuan on each violation. It was ignored. I then increased the penalty to 20 Yuan, and it was ineffective until I imposed a penalty of 50 Yuan. They feel that they gain 50 Yuan if they are not caught after spitting. They feel that they gain if they are not caught when they are not working. They are greedy. If I can find 300 Hong Kong workers of 20 years old, I would rather manufacture in Hong Kong. (personal interview, August 1994)

The problems of factory fires,[10] inhuman living conditions, low wages and physical abuse in small- to medium-sized Hong Kong and Taiwanese companies in Shenzhen and the Pearl River Delta areas have caught the attention of foreign media (including Hong Kong media; Clifton, 1994; Goldstein & Huus, 1994;). Fire accidents are less frequent in electronics factories than in factories producing plastics, toys, textiles, or shoes, because materials in electronics factories are mostly inflammable. Furthermore, the working environment in electronics is less polluted in terms of air and noise (Li, 1994). The workplaces of the four companies I visited are relatively spacious and clean. However, there are safety issues as many management personnel lock fire escapes to prevent workers from stealing products, and the popular "three-in-one" facility (factory, storage, and dormitory) is life-threatening when fire breaks out in any of the units.

---

[10]The Zhili fire disaster aroused concerns about foreign investors' neglect of industrial safety in the SEZs. More than 80 workers were burnt to death in the fire accident at the Zhili Toy and Crafts factory, a Hong Kong venture, on November 19, 1994. Another toy factory, built without authorization on a Shenzhen industrial site, collapsed June 4, 1994 from shoddy construction, killing at least 11 workers and injuring 27 others. In Zhuhai SEZ, next to Macau, 76 workers, who were sent back into a burned out textile factory to salvage materials, died when the building collapsed in June 1994.

Migrant workers employed in the Shenzhen SEZ face a situation different from that of the early 1980s, when most workers recruited by the joint ventures had previous production experience in state-run enterprises. Many of them had contracts and associated social security—no matter how limited they were—in the foreign-financed factories. Moreover, the establishment of labor unions was officially encouraged, even though the unions might have functioned more to facilitate management control of foreign investors than to fight for worker welfare. However, there were tradeoffs between economic benefits and political status. In a *Modern Life* (a Guangdong daily) 1986 survey of 900 workers in joint ventures with foreign corporations, the respondents expressed satisfaction with relatively high wages but also felt losses in status working for "capitalist enterprises" as well as the disappearance of life-long employment, fixed wages, and broad-based living subsidies and welfare benefits (cited in Leung, 1988). Leung quoted a worker as saying:

> I still don't understand why we have to enter into joint ventures with bosses and that the masters of our country, the workers, have to work for capitalists. Are we really that poor that we can't even afford this little bit of dignity? But then again, it's true that in these several years, all my colleagues and my living has got much better. Now everybody's got TVs, fridges, hi-fis and washing machines. It is all the result of joint ventures. (p. 125)

In recent years, the workers recruited were usually unemployed in their home villages. They never had job stability before, and many are happy with the wages they receive, especially those paid in Hong Kong dollars. They migrate to the SEZ in an era when capitalist ideology prevails, and, therefore, they are no longer treated as "bad elements" for working in foreign-funded enterprises.

However, workers are offered neither contracts nor government-funded social amenities, and they do not receive medical and many other basic employment benefits. They are discouraged from marrying and establishing families in the SEZ; if they do, they are required to pay penalties such as higher school fees. Their only hope is to work long hours, seven days a week, and then, years later, to return to their home villages with some savings (Cheng & Mosher, 1992).

Shenzhen recorded 1,100 cases of collective labor disputes from August 1992 to August 1994, with most taking place in foreign-owned factories, even though, overall, factories operated by Western companies tend to have better working conditions and labor-management relations than factories of Asian companies.

Shenzhen authorities have been worried about capital flight if industrial peace is not maintained and word spreads that Chinese workers are no longer "docile" and "hard working." At the 1987 Sanyo plant incident, strike leaders were fired not by Sanyo but by Chinese authorities, and each of the workers was fined (Leung, 1988). Already some Taiwanese capital has been diverted to Vietnam ("Clean Clothes," 1995, p. 6).

WORKERS IN THE CHANGING REGIONAL/ INTERNATIONAL DIVISION OF LABOR

Advances in microelectronics and automation technology may have rendered cost-cutting strategies by offshore production less significant for TNCs. Yet that does not imply that TNCs are moving onshore, and that workers in NICs are no longer integrated into the global assembly line. The ANICs have increasingly been capable of producing low- and medium-end I/CT products. TNCs are not giving up these product lines; instead, they are integrating countries into their global and regional subcontracting system, while retaining firm control of marketing and distribution channels and the supply of critical components. ANIC capital manufactures for TNCs increasingly on an OEM basis on the soil of their aspiring NIC neighbors and exercises management control from their home bases. Whereas TNCs deploy automated technologies and perform higher value-added functions in their home countries, the ANICs have set up manual operations in the special economic zones in China and export-processing zones in Southeast Asia.

In the Hong Kong-southern China region, the manufacturing done by Hong Kong for decades has been largely taken over by Shenzhen. Hong Kong has shifted to strategic planning, design and product development, finance management, marketing, and branch plant management. Manufacturing workers are forced to compete for jobs with their counterparts in southern China.

Shenzhen, however, is increasingly losing its competitiveness in attracting production investment to the Pearl River Delta region, which offers cheaper labor, land, and more tax incentives. The southern Chinese regional division of labor is structured in a way that Shenzhen and the Delta cities have become similar but separate entities dominated by Hong Kong capital, and they compete against each other for foreign direct investments and industrial subcontracting (Chan, 1991). Nevertheless, the rising costs of land, living, and labor in southern Guangdong will continue to drive Hong Kong manufacturers northwards. Eventually, Hong Kong managers will not be able to commute to Hong Kong in a day or two, and coordination and transportation costs will increase.

Hong Kong manufacturers' neglect of the welfare of workers in China and their abuses of laws and regulations are likely to induce a backlash from Chinese authorities and potentially "Hong Kong bashing." Although Guangdong has been closely tied to Hong Kong and tends to accept the Hong Kong way of doing business, the inland areas may not. The animosity toward "Hong Kong Chinese" may not manifest itself in the same way, however, that it does against affluent Chinese in Malaysia and Indonesia. With Hong Kong becoming part of China in 1997, and Hong Kong industrialists looking less like foreign investors and more like national entrepreneurs, the polarization is likely to turn on domestic class and regional inequalities.

It is possible that workers in China will rise up to demand a greater share of the economic growth. International labor organizations and human rights watchers are already pressuring the Chinese government and corporations to observe workers' rights. The continuation of the present form of labor management will provide an excuse for Western countries to limit the import of products manufactured in China. The existing production structure, labor recruitment, and management practices will have to change to reduce the social and economic inequities that are emerging.

ACKNOWLEDGEMENTS

The author wishes to thank Professor Richard Child Hill and Professor Bella Mody for guiding the dissertation on which this article is based, and the editors for editing the chapter.

REFERENCES

Aldana, C.H. (1989). *A contact for underdevelopment: Subcontracting for multinationals in the Philippine semiconductor and garment industries.* Manila, Philippines: IBON Databank Phils.

Andors, P. (1988). Women and work. *Bulletin of Concerned Asian Scholars,* 20(3), 22-24.

Chan, T.M.H. (1991). Economic development in the Shenzhen special economic zone: Appendage to Hong Kong?" *Southeast Asian Journal of Social Science,* 19(1 & 2), 180-205.

Chen, D. (1991). *Annual departmental report 1991.* Hong Kong: Government Printer.

Chen, X.M. (1987). Magic and myth of migration: A case study of a special economic zone in China. *Asia-Pacific Population Journal,* 2(3), 57-86.

Cheng, E., & Mosher, S. (1992, May). Economic strategy inspired by Hong Kong. *Far East Economic Review*, pp. 26-30.

Clean Clothes. (1995, March). Shenzhen newsletter, p. 6.

Clifton, T. (1994, December). Asia's fatal factories. *Newsweek*, pp. 20-23.

Eisold, E. (1982). *Young women workers in export industries: The case of the semiconductor industry in Southeast Asia* (Unpublished manuscript). Geneva, Switzerland: International Labor Organization.

Elson, D. (1981). *Women workers in export-oriented industries in Southeast Asia: A select annotated bibliography*. Brighton, UK: Institute of Development Studies.

Ernst, D., & O'Conner, D. (1992). *Competing in the electronics industry: The experience of newly industrializing economies*. Paris: Organisation for Economic Cooperation and Development Development Center.

Frank, A.G. (1967). *Capitalism and underdevelopment in Latin America: Historical studies of Chile and Brazil*. New York: Monthly Review Press.

Frank, A.G. (1981). *Crisis: In the Third World*. New York: Holmes & Meier.

Friedman, A.L. (1977). *Industry and labor. Class struggle at work and monopoly capitalism*. London: Macmillan.

Friedmann, J., & Wolff, G. (1982). World city formation: An agenda for research and action. *International Journal of Urban and Regional Research, 6*(3), 309-343.

Fröbel, F., Heinrichs, J., & Kreye, O. (1980). *The new international division of labor*. Cambridge, UK: Cambridge University Press.

Fuentes, A., & Ehrenreich, B. (1984). *Women in the global factory*. Boston, MA: Institute for New Communications, South End Press.

Goldstein, C., & Huus, K. (1994, June, 16). No workers' paradise: Labor activists make little headway in Shenzhen. *Far Eastern Economic Review*, pp. 35-36.

Grossman, R. (1979). Women's place in the integrated circuit. *South-East Asia Chronicle, 66*, 2-17.

Henderson, J. (1989). *The globalization of high tech production: Society, space, and semiconductors in the restructuring of the modern world*. London: Routledge.

Heyzer, N. (1986). *Working women in South-East Asia: Development, subordination and emancipation*. Exeter, UK: Open University Press.

Ho, Y.P. (1992). *Trade, industrial restructuring and development in Hong Kong*. Honolulu: University of Hawaii Press.

Hollander, N.C. (1977). Women workers and the class struggle: The case of Argentina. *Latin American Perspectives, 4*(1-2), 180-193.

Hong Kong Economics Journal. (1994, January). Special feature, p. 40.

Hong Kong Electronics Association (HKEA). (1994). *Directory of Hong Kong electronics industry*. Hong Kong: Author.

Hong Kong Government Industry Department (HKGID). (1991). *Techno-Economic and Market Research Study on Hong Kong's Electronics Industry, 1988-1989* (research conducted by Dataquest for Hong Kong Government Industry Department). Hong Kong: Author.

Hong Kong Government Industry Department (HKGID). (1993). *Hong Kong's Industries 1993*. Hong Kong: Author.

Hong Kong Trade Development Council (HKTDC). (1991, April). *OEM business with Japan's electronics industry*. Hong Kong: Author.

Hong Kong Yearbook 1994. (1994). Hong Kong: Government of Hong Kong.

Hong Kong Yearbook 1996. (1996). Hong Kong: Government of Hong Kong.

Hymer, S. (1972). The multinational corporation and the law of uneven development. In J.W. Bhagwatti (Ed.), *Economics and world order* (pp. 113-140). New York: MacMillan.

Katz, N., & Kemnitzer, D.S. (1983). Fast forward: The internationalization of Silicon Valley. In C. Jun & M.P. Fernandez-Kelly (Eds.), *Women, men, and the international division of labor* (pp. 332-345). Albany: State University of New York Press.

Leung, W.Y. (1988). *Smashing the iron rice pot: Workers and unions in China's market socialism*. Hong Kong: Asian Monitor Resource Center.

Li, N.G. (1994, April). The labor exploitation of foreign investors. *The Nineties*, pp. 53-55.

Mody, A., & Wheeler, D. (1990). *Automation and world competition*. New York: St. Martin's Press.

Ng, I.W.C. (1992). *Flexible production and the creation of competitive advantage in an Asian newly industrializing economy*. Unpublished doctoral dissertation, University of California, Los Angeles.

Overholt, W. (1993). *China: The next superpower*. London: Weidenfeld & Nicolson.

Piore, M., & Sabel, C. (1984). *The second industrial divide: Possibilities and prosperity*. New York: Basic Books.

Rada, J. (1980). *The impact of microelectronics: A tentative appraisal of information technology*. Geneva, Switzerland: International Labour Organisation.

Sanderson, S.W., Williams, G., Ballenger, T., & Berry, B.J.L. (1989). Impacts of computer-aided manufacturing on offshore assembly and future manufacturing locations. *Regional Studies, 21*(2), 131-142.

Sassen-Koob, S. (1980). The internationalization of the labor force. *Studies in Comparative International Development, 15*.

Sayer, A. (1986). New development in manufacturing: The just-in-time system. *Capital and Class*, No. 30, 43-72.

Scott, A.J. (1983). Location and linkage systems: A survey and reassessment. *Regional Science, 17*(1), 1-39.

Sivanandan, A. (1979). Imperialism and disorganic development in the silicon age. *Race and Class, XXI*(2), 111-126.

Sklair, L. (1991). Problems of socialist development: The significance of Shenzhen special economic zone for China's open door development strategy. *International Journal of Urban and Regional Research, 15*(4), 197-215.

Smart, J., & Smart, A. (1991). "Personal relations and divergent economies: A case study of Hong Kong investment in South China. *International Journal of Urban and Regional Research, 15*(4), 216-233.

United Nations Centre on Transnational Corporations UNCTC. (1987). *Transnational corporations and the electronics industries of ASEAN economies.* New York: United Nations.

United Nations Economic and Social Commission for Asia and the Pacific (UNESCAP). (1987). *Economic and social survey of Asia and the Pacific.* Bangkok: Author.

Vennin, B., & de Banville, E. (1975). Pratique et signification de la sous-traitance dans l'industrie automobile en France. *Review Economics, 26*(2), 280-306.

Wallerstein, I. (1974). *The modern world-system: Capitalist agriculture and the origins of the world-capitalist economy in the sixteenth century* (2 vols.). New York: Academic Press.

Wallerstein, I. (1979). *The capitalist world-economy.* Cambridge, UK: Cambridge University Press.

Watanabe, M. (1993, March/April). Some thoughts on technology transfer. *JETRO,* No. 103.

Yamada, B. (1990). *Internationalization strategies of Japanese electronics companies: Implications for Asian newly industrializing economies* (Tech. Paper No. 28). Paris: Organisation for Economic Cooperation and Development.

Yu, J.M. (1996, April). Labor exploitation by foreign investors. *The Nineties,* pp. 56-59.

8

Challenges to Hollywood's Labor Force in the 1990s

Janet Wasko

Commercial filmmaking in the United States, centered in the Los Angeles area or "Hollywood," has been highly unionized. Early in the industry's history, film workers were organized by trade unions from related industries such as the theater and the electrical industry. Eventually unions and guilds were formed specifically to organize Hollywood workers, and most of these labor groups are still active in the film and television industries. Similar to other U.S. labor organizations, the Hollywood unions and guilds continue to be challenged by political and economic developments in society in general and the film industry in particular.

The global expansion of the film industry over the last few decades has proven to be especially problematic for U.S. film workers and presented problems for cinema workers in other parts of the world. There is no doubt that the global business for films and other entertainment products has grown dramatically in the last decade. For instance, worldwide revenues for theatrical rentals, home video, and television for U.S. companies in 1991 exceeded $13 billion (see Wasko, 1994a). However, a political economic analysis focusing on labor issues presents a gloomy picture for trade unions and workers in the entertainment business. Although some top stars, writers, and directors are benefitting heavily from these developments, a more careful look at the power relations in Hollywood reveals a different picture for many other workers. This chapter discusses current issues facing Hollywood's labor organizations and workers in the 1990s, including the increase in nonunion production and the diversification and internationalization of the entertainment industry.

HISTORICAL BACKGROUND/CURRENT STATUS OF HOLLYWOOD LABOR ORGANIZATIONS

According to data from the Bureau of Labor Statistics, motion picture workers in the U.S. totaled 404,000 in 1992 and 421,000 in 1993. Film workers are a highly skilled and specialized labor force. Unemployment is high. For instance, 85% of actors are out of work most of the time (Raskin, 1988). There are some unusual or unique characteristics as well. Some workers such as writers, directors, and actors share in the profits of films through profit participation deals. Others may become employers themselves through their own independent production companies or in projects in which they serve as producer or director e.g., Billy Crystal worked as an actor in "City Slickers II" but also was the film's producer). There also are keen differences between "creative" (above-the-line) and "craft" (below-the-line) workers, with consequent differences between the labor organizations that represent these different types of labor (Nielsen, 1985). In other words, the organization of entertainment unions along craft lines rather than a vertical, industrial structure has tended to inhibit labor unity within the industry.

The history of Hollywood labor organizations, although neglected for many years by scholars, has received increased attention during the last decade (Clark, 1989; Nielsen, 1985; Prindle, 1988; Staiger, 1981). However, most of these excellent studies have focused on the historical background of Hollywood labor, with less attention given to more recent developments such as globalization. Only a brief introduction to the major trade organizations is given here, followed by a discussion of current issues.(See Table 8.1 for a summary of the main labor organizations and their membership figures.)

Table 8.1. Trade Unions Active in the U.S. Film Industry.

Above-the-Line Organizations (Professional/Performers)	Founded	Membership	
		1987	1993
Directors Guild of America (DGA)	1937	7,751	9,500
Screen Actors Guild (SAG)	1936	70,000	75,000
Screen Extras Guild (SEG)	1946	4,000	5,300
Writers Guild of America (WGA)	1954	9,030	9,600
Below-the-Line Organizations (Craft/Technical)			
International Alliance of Theatrical and State Employees & Motion Picture Operators (IATSE)	1893	58,500	65,000
National Association of Broadcast Employees and Technicians (NABET)[a]	1933	5,000	20,000
International Brotherhood of Teamsters Local 399—Studio Transportation Drivers	1903	2,300	2,900[b]

Sources: Data compiled from *Encyclopedia of Associations* (1992); Kleingartner and Raymond (1988).

[a]NABET became part of the Communication Workers of America (CWA) in January 1994.
[b]Estimate based on information received from Local 399 official.

BELOW-THE-LINE UNIONS[1]

The International Association of Theatrical and Stage Employees (IATSE or IA) has been the most powerful union active in the U.S. film industry. Formed at the end of the 19th century, IATSE organized stage employees in the United States and Canada (Ross, 1941). As the entertainment industry expanded, IATSE grew to include motion picture projectionists and technical workers at the Hollywood studios and film exchanges throughout North America. When television was introduced, IATSE organized technical workers in the new medium.

IATSE has a tradition of local autonomy, with a variety of craft-based locals involved in collective bargaining agreements. However, nationwide agreements for film production personnel are negotiated with the producers association—the Alliance of Motion Picture & Television Producers (AMPTP).

[1]Background for the unions and guilds discussed has been based on a variety of sources, but especially descriptions in Fink (1977).

IATSE's history includes some dismal chapters from the 1930s when racketeers and criminals extorted funds from union members, as well as others from the 1940s when the union assisted in the ugly blacklisting activities that tainted Hollywood (Nielsen, 1985).

The National Association for Broadcast Employees and Technicians (NABET) grew first out of radio and then television broadcasting. The union was organized at the National Broadcasting Corporation (NBC) as a company union (an industrial, rather than craft-oriented organization) as an alternative to the larger and more powerful IBEW (Koenig, 1970). NABET's relatively militant history is replete with skirmishes with IBEW and IATSE, along with continuous rumors of a merger with the larger IATSE (Wasko, 1983).

In 1990, NABET's Local 15, which organized 1,500 freelance film and tape technicians in New York, merged with IATSE ("NABET Local 15 votes," 1990; Miller, 1990) Then, in 1992, most of the other NABET locals agreed to join the Communication Workers of America (CWA), effective January 1994. About 9,300 NABET members became a part of the much larger CWA, which represents 600,000 workers in telecommunications, printing, broadcasting, health care, and the public sector. A CWA official explained the merger in straightforward terms, highlighting labor's current concerns with unity and globalization:

> In this day and age, with all the concentration of corporate power, it has become an advantage for unions to band together and join their resources and strength. It certainly helps when unions have to take on these multinational corporate structures, as especially evidenced in the communications and broadcasting fields. (O'Steen, 1992a)

Although most of NABET's members were to be moved to an independent broadcasting arm in CWA, NABET's West Coast Local 531 agreed to merge with IATSE because of its 500 members' closer affiliation with the film industry (O'Steen, 1992b). Thus, IATSE became the only union in the United States. to represent behind-the-camera film workers. About the same time, NABET also held merger talks with several Canadian media unions (Papp, 1992).

The International Brotherhood of Teamsters is the largest and strongest union in the United States and also active in the motion picture industry, organizing studio transportation workers on the West Coast and various other workers. The Teamsters claimed a general membership of over 2 million in 1986; its Hollywood Local 399 has approximately 2,900 members who work as truck drivers and security personnel in the film industry. (The Teamsters also have organized workers involved with other aspects of corporate Hollywood's activities, as discussed later.)

ABOVE-THE-LINE GUILDS

The Screen Actors Guild (SAG) was organized in 1933, after several other organizations had attempted to organize film performers, including the Academy of Motion Picture Arts and Sciences (Clark, 1989; Prindle, 1988).

The history of SAG is at first dominated by the attempt to establish a guild shop and works now to gain compensation for actors in the constantly expanding forms of distribution (television, video cassettes, etc.). SAG's concern with compensation is not an insignificant issue considering that its members gained more than $1 billion in 1987 merely from residual payments for TV reruns of old films (Raskin, 1988).

In 1992, the Screen Extras Guild's (SEG) 3,600 members became a part of SAG's union coverage, primarily because SEG lacked the clout to deal with producers and most extras were working non-union members (Stumer, 1992).

Serious discussions of a merger between SAG and the American Federation of Television and Radio Artists (AFTRA) have taken place over the last several years (O'Steen, 1992c). AFTRA was formed in 1937 to represent radio and then television performers. The organization's primary jurisdiction is in live television, but AFTRA shares jurisdiction with SAG for taped television productions. AFTRA's 75,000 members also include some vocalists (Stumer, 1992). The merger still may take place, but negotiations have been delayed because of a major law suit involving AFTRA's Health and Retirement Fund (Robb, 1994).

The Writers Guild of America (WGA) represents movie and TV writers but is split between WGA East (3,500 members) and WGA West (6,500 members). The organization was founded in 1954, although writers were organized previously by groups such as the Authors League of America and the Screen Writers Guild (Fink, 1977; Schwartz, 1982). Although a few blockbuster writers have made deals for $1 million film scripts over the last few years, many writers have experienced the same problems as other workers: lower pay rates and fewer jobs due to recession, and so on.

The Directors Guild of America (DGA) represents directors, unit production managers, assistant directors, and technical coordinators in television and film. The guild was formed in 1960 from the merger of the Screen Directors Guild and the Radio and Television Directors Guild. Its organizational membership was about 9,700 in 1992. Prompted especially by the introduction of colorized films, the DGA has lobbied strongly for a moral rights law for creative personnel to prevent changes in their work (Stumer, 1992).

The American Federation of Musicians (AFM) represents musicians who work in the film industry. The trade group, which was formed in the 1890s, has negotiated contracts with the industry since 1944 and has been especially concerned with new technological developments in sound recording (Koenig, 1970; Leiter, 1953).

ISSUE 1: NONUNION AND NON-HOLLYWOOD PRODUCTION— WHICH SIDE ARE YOU ON?

The biggest headache facing these Hollywood unions and guilds is the proliferation of nonunion production in the Los Angeles area as well as at production sites all over the country and the world.

Film and television production around Los Angeles seems to ebb and flow, depending on a number of different factors, including the lure of cheaper locations. One commonly used indicator of filming activity has been the number of permits issued by the Los Angeles Film Office. Table 8.2 indicates a decline in the number of permits issued since 1990, which again may be attributed to a number of factors including changes in TV programming, and so on. (Beving, personal communication, June 6, 1994).

There are differing reports of the prevalence of nonunion production in Hollywood. According to one source, more than 90% of the permits issued by Los Angeles for film work in 1979 went to

Table 8.2. Film Permits Issued by Los Angeles City Film Office, 1984-1993.

	Features	TV	Comm'ls	Music Videos	TOTAL[a]
1984	532	1,522	911	191	3,511
1985	519	1,575	921	141	3,488
1986	912	1,467	1,197	151	4,056
1987	888	1,370	1,181	216	3,968
1988	930	942	1,242	225	3,695
1989	1,110	1,627	1,434	310	4,815
1990	1,192	1,879	1,436	304	5,146
1991	1,074	1,545	1,223	330	4,567
1992	931	1,463	1,171	328	4,231
1993	957	1,105	1,180	353	3,914

Source: Los Angeles Film & Video Permit Office.

[a]Includes permits issued for miscellaneous productions.

companies that hired only workers covered by union contracts. By 1986, only 60% of the permits went to fully unionized major studios, and in January 1989 only 40% of permits were issued for unionized productions (Bernstein, 1989b).

According to IATSE reports, 65% of films produced in southern California in 1989 were made with nonunion crews (Stremfel, 1989). More recently, IATSE has claimed that only 30% (121 out of 400) of the pictures released in the U.S. in 1993 were made with union labor; in 1992, the figure was 109 out of 390 films (28%; Cox, 1994b).

Why the move toward nonunion workers and new locations? The answer is multifaceted. As always, employers are trying to lower labor costs, and there is a ready supply of nonunion workers, both in Hollywood and other locations. In addition, the established entertainment unions are perceived as uncooperative and too demanding.

The abundance of available labor also may be related to the popularity of media in general. The growth of media education at universities and colleges, as well as the increased visibility of film and television production in the popular press, means that there is a glut of eager workers for Hollywood companies to employ, very often, without union affiliation. For example, over 3,000 applications are submitted to the DGA's Assistant Directors Training Program each year—but only 12 are chosen (Marx, 1994).

In addition, because of the fantasy associated with Hollywood, even "regular" work in the film industry seems glamorous, as indicated by the following explanation from a back lot laborer, writing about his job:

> The motion picture and television industry also has its laborers. . . . The laborer (officially referred to as a Studio Utility Employee) is the lowest paid craftsman in the business. On the other hand, there are certain compensations as compared to being a laborer in other industries. He works in the middle of a glamour industry, in studios, surrounded by stars, starlets and props, and he gets to see how things are done. What is it like to be in this fantastic world where even a sweating laborer covered with dust (not star dust) or dirt from a ditch he is digging might look handsome enough to be the leading man in a picture? (Many would-be actors have ended up in the studio crafts and that is most likely why Hollywood has the best looking [as a group] craftsmen of any industry.) (quoted in Levenson, 1972, p. 5)

While studios try to blame unreasonable union demands for the increase of nonunion production and the flight to nonunion locations, labor leaders (especially from below-the-line unions) claim that they are not the problem. They point to the skyrocketing costs of above-the-line talent, with especially high salaries going to high-profile actors and actresses. As one union official explains, "Until they can control their above-the-line costs and their own studio's executives, they'll never bring costs back in line. They can beat us until we do it for free, but if Julia Roberts still wants 76 million bucks, the picture is still going to cost" (Cox, 1994b, p.1).

The lack of unity among entertainment unions also has been blamed for the growth of nonunion filming. Some of the mergers mentioned previously may help to alleviate this problem, yet the organization of labor along craft lines still exacerbates the situation (Miller, 1989).

The lure of lower budgets with nonunion workers has attracted producers to right-to-work states such as Florida, as well as other states that have recognized film and television production as a boost to local economies. (See Table 8.3 for state employment figures.) Local communities are said to receive about 40% of funds from productions shooting in their neighborhoods (Sumner, 1993). Meanwhile, foreign locations such as Eastern Europe and parts of the Third World offer low budgets and exotic locations.

With the flight to nonunion states and the attraction of other locales, some have argued that the film industry is now geographically fragmented or characterized by "flexible specialization." From their base in urban planning and drawing on neoclassical economics, Storper and Christopherson (1987) argue that the film industry has been restructured from the integrated, mass-production studio system of the 1930s and 1940s (the Fordist model) to a disintegrated and flexible system based on independent and specialized production (the post-Fordist model). Thus, the film industry provides an example of the viability of flexible specialization for other industrial sectors to emulate.

Table 8.3. Film and Television Employees by State (1992).

	No. of employees	% National Total
California	79,000	67.0
New York	11,500	9.7
Illinois	3,500	3.0
Florida	2,000	1.7
Texas	1,900	1.6
National Total	118,000	

Source: New York City Economic Policy and Marketing Group.

Although these interpretations describe some important changes in the U.S. film industry of the late 20th century, the analysis is severely handicapped by the emphasis on production and the neglect of the key roles played by distribution, exhibition, and financing. Asu Aksoy and Kevin Robins (1992) provided an excellent critique of the flexible specialization thesis: "For them [Storper and Christopherson] the major transformation in the American film industry is centered around the reorganization of production, and, more particularly, around the changing relationship between technical and social divisions of labor in production. It is as if the Hollywood industrial story begins and ends with the production of films" (p. 10). The flexible specialization argument also overlooked the considerable concentration of postproduction in California. However, this may be changing, as some industry spokesmen have observed growing postproduction activities in Florida and Vancouver, Canada (O'Steen, 1994).

Pressure from the availability of a nonunion option and runaway production has forced the unions to make concessions during contract negotiations. An example is IATSE's recent contract negotiation process with the AMPTP, when a strike was barely averted with a new contract accepted in December 1993. The AMPTP claimed that they had "received the necessary concessions"—a side letter agreement easing restrictions on producers of TV movies, series, and pilots on such elements as wages, vacations, holidays, overtime, transportation allowance, and interchangeability of employees on the set. In addition, the IA agreed to "give good faith consideration" for low-budget features on a case-by-case basis (Welkos, 1993).[2]

The union, however, also claimed victory, receiving health and pension benefits, plus a mandate to study the "contentious" Article 20 clause of the contract. The clause addresses the practice of studios funding independent nonunion productions when there is no creative control, and it allows studios to pick up "bona fide" nonunion productions for distribution if the unions are notified. The studios claim that they are encouraging low-budget independent filmmaking, whereas union officials explain that their efforts are aimed at trying to keep jobs in Los Angeles.

As part of the contract negotiations, a committee, including IA and AMPTP representatives, was appointed to study the definition of a bona fide production-distribution deal (Cox, 1994a). Committee meetings became especially tense when a union report leaked to the

---

[2]The basic crafts unions also made concessions during their negotiations in April 1994. The unions include Teamsters Local 399, Laborers Local 724, Plasterers Local 755, and IBEW Local 40, which have bargained separately from IATSE since 1988 (see Robb, 1994a).

media claimed that films made under Article 20 had nearly tripled between 1991 and 1993, and that the studios were setting up dummy independent production companies to gain concessions and wage cuts, while the films are actually controlled by the studios. The studios countered by claiming that 60% to 70% of the work of AMPTP companies is unionized (however, with concessions; Cox, 1994b).

During these intense contract negotiations, some argued that IA has lost its clout. Even if the union went on strike, the producers and other observers felt that there was enough nonunion labor to continue production. They also assumed that IA members themselves would have worked nonunion jobs if a strike had been called (Cox, 1993).

Meanwhile, unions in New York also have been forced to accept concessions in order to lower labor costs and compete with Los Angeles. The East Coast Council—a coalition of seven motion picture and television production unions—has actively tried to lure work away from Hollywood, with substantial changes in contracts and concessions, including wage reductions of up to 50% (Mirabella, 1993; Span, 1992).

ISSUE 2: CORPORATE DIVERSIFICATION IN HOLLYWOOD— MICKEY AS TEAMSTER

"Hollywood" refers to that set of corporations that no longer simply produces and distributes motion pictures, but markets a wide range of entertainment products (e.g., television programs, video cassettes, and music products), as well as operating theme parks and owning professional sports teams (Wasko, 1994b). Although Hollywood companies have always participated in a variety of media-related activities, these companies are even further diversified today than in the past.

Union representation also has become more diversified, as the different types of businesses incorporated by Hollywood companies have involved further differentiation of labor, making it difficult for workers to form a united front against one corporation. For instance, workers employed by Disney include animators at the Disney Studio; hockey players on Disney's hockey team, the Mighty Ducks; and Jungle Cruise operators at Disney's various theme parks.

The differentiation of labor is especially apparent at the theme parks owned by many Hollywood companies, in particular, Disney, Universal, Paramount, and Time Warner. Workers at these sites are represented by a wide array of labor organizations, many of which are unrelated to those unions active in the film industry.[3] For instance, over

[3]Amusement and recreation workers in the United States far outnumber those in the film industry: in 1992, 1,169,000 and in 1993, 1,181,000.

a dozen unions have contracts with Disney World ("Actors' Equity Concludes," 1990). Meanwhile, at Disneyland, five unions usually negotiate a master agreement for about 3,000 employees. The trade unions include the United Food and Commercial Workers, Service Employees International Union, Hotel Employees and Restaurant Employees, Bakery, Tobacco and Confectionery Workers, and the Teamsters (who represent workers who wear the life-size costumes of Disney characters at the park).

Although the notion of "Mickey Mouse" and "Donald Duck" as Teamsters may be jolting to many Disney aficionados, at least they are represented by an employee association. The same is not true for characters such as "King Kong" and "E.T." at Universal's theme park in Florida. Universal's Studio Tour in California is unionized, but the company's studio/theme park in Florida is totally nonunion. Although several unions have attempted to organize workers, Universal has conducted what some union officials have labeled "viciously antiunion campaigns," far outspending the unions in convincing employees to vote against unionization. One campaign included an employee questionnaire and raffle for various prizes, a practice questioned by the National Labor Relations Board. As one of Universal"s administrators explained, "We're a young growing company in a tourist-based environment. The union structure does not add value to the work environment or the guest environment" ("Universal Studios," 1993, p. A8).

Generally, then, the trend toward diversification has contributed to a weakening of trade unions' power as well as a further lack of unity among workers. As *Los Angeles Times* labor reporter, Harry Bernstein (1989a), observed, "These days, corporate tycoons own conglomerates that include businesses other than studios and networks. They may enjoy movie making, but money seems to be their primary goal. So if production is stopped by a film industry strike, their income may be slowed, but money can still roll in from other sources."

ISSUE 3: INTERNATIONALIZATION OF THE ENTERTAINMENT BUSINESS—ENTERTAINMENT WORKERS OF THE WORLD UNITED?

International corporate expansion has long been of vital concern to workers organizations, as indicated by the formation of international trade secretariats at the end of the last century.

Globalization in the entertainment industry has intensified in recent years, as diversified media conglomerates such as Time Warner distribute their products to new channels and newly opened markets around the world. Indeed, the mergers that create a large and dominant corporation such as Time Warner are justified because of the importance of competing internationally (Barnet & Cavanagh, 1994; Time Warner, 1992).

Entertainment unions also have formed a number of international organizations, although they have had unstable histories, and any real influence such as international collective bargaining agreements is rare.

One example of such collective activity among international labor organizations is the International Federations of Performers (FFF),[4] formed by the International Federation of Musicians (FIM) and the International Federation of Actors (FIA).[5] The FFF negotiates collective agreements with employers' groups such as the European Broadcasting Union (EBU) on international television relays and exchange of radio broadcasts, the International Federation of the Phonographic Industry (IFPI) on participation in broadcasting revenue, and the International Radio and Television Organization (OIRT). As of 1980, FFF was the only international organization engaged in collective bargaining, with its primary concern being the effects of technological change within the industry. Its policy was to obtain, whenever possible, more control over and compensation for repeated use of performances by union members, and remuneration for displaced performers. In a 1980 article describing FFF's agreements, Miscimarra (1980) found "multinational entertainment union activity to be advanced beyond that in other industries" (p. 59) and that there was "a degree of international union solidarity which is completely unprecedented" (p. 59).

---

[4]The history of international labor secretariats in the entertainment and media field is quite complex, and much more attention should be directed at these organizations in future research. Currently, the two major groups are: (a) the International Federation of Unions of Audio-Visual Workers (FISTAV), founded in 1974, and the International Secretariat of Entertainment Trade Unions (ISETU), founded in 1965 by the International Confederation of Free Trade Unions (ICFTU), merged in 1993 to form ISETU/FISTAV—its members include 112 unions in 50 countries, affiliated for 200,000 members; and (b) the International Committee of Entertainment and Media Unions (ICEMU), formed by the International Federation of Journalists (IFJ) in 1992 and which includes FIM, FIA, ISETU/FISTAV, International Graphical Workers Federation (IGF), and European Federation of Audiovisual Film Makers (FERA). Background on these organizations is from material provided by the groups, plus Encyclopedia of American Associations (1992) and Rowan, Pitterle, and Miscimarra (1983).

[5]The FIM was founded in 1948 to discuss the dangers of mechanical music and other forms of technological change and in 1991 included national unions totaling 30,000 musicians in 19 countries. The FIA was founded in 1952 with affiliated unions in 43 countries.

Miscimarra pointed to several reasons why there should be more cooperation between workers and labor organizations internationally. The international nature of filmmaking and the effects of technological change on employment conditions should serve as positive motivations for union cooperation.

However, some of the basic characteristics of entertainment industry labor also complicate these collective efforts. As Miscimarra (1980) observed: "The industry is marked by a large number of different unions, a high degree of individual bargaining, a predominance of freelance, short-term employment, and few objective standards by which employees can be evaluated" (p. 60). As usual, there is fierce competition among unions and nations. Lower costs are offered in some countries (especially Eastern Europe and some Third World countries), whereas other countries publicize cooperative, rather than troublesome, workers.

These difficulties are exacerbated by the ideological and positional differences that union organizations sometimes take on in international trade issues and cultural policies.[6] Excellent examples of such contradictory behavior are seen in the positions U.S. unions have taken in support of efforts to enhance the distribution of U.S. films internationally, the policies behind such expansion, and the responses of European and other unions outside the United States.

One example is the recent discussion of international trade policies in negotiations over the General Agreement on Trade and Tariffs (GATT). From the beginning of its talks in Uruguay in 1986, the United States vowed to make sure that cultural products were included in the agreement, thereby providing the basis for eliminating quotas and subsidies that restrict Hollywood products in foreign markets (especially in Europe). However, cultural products were excluded when the United States agreed to compromise in order to save the whole agreement (see Wasko, 1994a, 1994c).

Hollywood has always taken a strong position against protective measures, sometimes fighting with a united front of producers as well as trade unions. Leaders of 10 major entertainment unions joined the industry trade associations (the MPAA and AFMA) in rallying against cultural exemptions in GATT (Harris, 1992). AFMA president Jonas Rosenfield explained that the situation was "historic in that it represents a cross-section of labor and management in the industry" (quoted in Frook, 1992, p. 1). In a letter of support, the unions reminded U.S. trade representatives that the film industry represents "one of our most competitive and successful industries in world trade" and the nation's leading net exporter (Harris, 1992, p. 1).

---

[6]The U.S. unions and guilds joined some of these international organizations at a relatively late stage because of ideological reservations (see Miscimarra, 1981).

Meanwhile, the International Federation of Actors' (FIA) European group joined with European industry associations to keep the audiovisual sector outside of GATT, arguing that any further opening would reinforce the U.S. dominance of the European market ("EC U.S. differences remain," 1993). Thus, union organizations found themselves on opposite sides of the battle over international trade policies.

The North American Free Trade Agreement (NAFTA) presented another example of international trade policy differences among unions. Many corporate executives in Hollywood were ecstatic about NAFTA, envisioning benefits in consumer, exhibition, and distribution markets, especially in Mexico, as well as an improvement in curbing film and video piracy. U.S. labor was one of the principal opponents of NAFTA—except Hollywood's unions, which, with the exception of IATSE, did not take a stance (Ulmer, 1993). IATSE became especially concerned about a rumor in July 1993 that the agreement would create a major new film studio in Mexico (Robb, 1993).

During its 15th Congress in Montreal, the FIA (including the Screen Actors Guild), although not supporting NAFTA, made a strong statement that supported cultural integrity of nations:

> The fundamental objective of FIA is the protection of the artistic, economic and social interests of performing artists within its member countries and unions, and whereas it has always been the policy of FIA to encourage the preservation and growth of each and every nation's cultural identity, be it resolved, that FIA condemns any attempt by any government, organization or individual, through the use of trade agreements, practices or policies, to undermine the integrity of a national culture, and the status of performing artists in those cultures. (Robb, 1992)

As Miscimarra (1980) concluded, "the national barriers to absolute union co-operation in the entertainment field will be slow to fall" (p. 59).

CONCLUSION

The pressures are mounting on labor organizations in the entertainment field. Hollywood unions and guilds have faced difficult struggles in the past, combatting a range of challenges—from union recognition in the 1930s to ideological assaults and blacklisting in the 1940s and 1950s. In the 1990s, they face further struggles with antiunion efforts and with power moves by diversified corporations actively involved in international markets and job relocation. As the processes of concentration and internationalization continue to shape the entertainment and media industries, it remains to be seen when and if entertainment workers of the world will revitalize labor organizations, recognize their collective interests, and build an international solidarity that matches the efforts of the industry.

REFERENCES

Actors' equity concludes initial accord covering performers at Walt Disney World. (1990) *Daily Labor Report*, No. 166, p. A-7.

Aksoy, A., & Robins, K. (1992). Hollywood for the 21st century: Global competition for critical mass in image markets. *Cambridge Journal of Economics*, 16(1), 1-22.

Barnet, R. J., & Cavanagh, J. (1994). *Global dreams: Imperial corporations and the new world order*. New York: Simon & Schuster.

Bernstein, H. (1989a, April 11). Hollywood may take the drama out of settling disputes. *Los Angeles Times*, p.1.

Bernstein, H. (1989b, January 24). Hollywood's craft workers under pressure to take cuts. *Los Angeles Times*, p. 1.

Carnicas, P. (1994). Behind the numbers: New York production in national context. *SHOOT*, p. 7.

Clark, D. A. (1989). *Actors' labor and the politics of subjectivity: Hollywood in the 1930s*. Unpublished doctoral dissertation, University of Iowa, Ames, IA.

Cox, D. (1993, December 20). Non-union clout puts producers in power post. *Variety*, p. 13.

Cox, D. (1994a, January 3). Happy new year for IA, producers. *Variety*, p. 15.

Cox, D. (1994b, May 18). IA hits "hidden" pix. *Daily Variety*, p. 1.

EC, U.S. differences remain on audiovisuals. (1993, October 20). *The International Trade Reporter*, p. 1777.

Encyclopedia of American Associations. (1992). Detroit, MI: Gale Research.

Fink, G. M. (Ed.). (1977). *Labor unions*. Westport, CT: Greenwood Press.

Frook, J. E. (1992, December 30). AFMA joins GATT quota battle. *Daily Variety*, p. 1.

Harris, P. (1992, December 29). Guilds push quota-free trade pact. *Daily Variety*, p. 1.

Kleingartner, A., & Raymond, A. (1988). *Hollywood goes international: Implications for labor relations*. Los Angeles: University of California, Los Angeles, Anderson Graduate School of Management.

Koenig, A. E. (Ed.). (1970). *Broadcasting and bargaining*. Madison: University of Wisconsin Press.

Leiter, R. D. (1953). *The musicians and Petrillo*. New York: Bookman Associates.

Levenson, J. (1972). *The back lot: Motion picture studio laborer's craft described by a Hollywood laborer*. Los Angeles: Levenson Press.

Marx, A. (1994, April 25). Hollywood wannabes are copping degrees. *Variety*, p. 1, 42.

Miller, R. (1989, September 1). NYC unions meet to discuss issue of non-union work. *Back Stage*, p. 1.

Miller, R. (1990, June 8). Will New York become a one-union film town?" *Back Stage*, p. 16B.

Mirabella, A. (1993, March 1). Labor helps end quiet on the set: A new attitude at film unions. *Crain's New York Business*, p.3.

Miscimarra, P. A. (1981). The entertainment industry: Inroads in multinational collective bargaining. *British Journal of Industrial Relations, XIX*, 49-65.

NABET Local 15 vote to merge into IATSE. (1990, September). *Entertainment Law Reporter*, p. 19.

Nielsen, M. (1985). *Motion picture craft workers and craft unions in Hollywood: The studio era, 1912-1948*. Unpublished doctoral dissertation, University of Illinois, Urbana.

O'Steen, K. (1992a, June 17). NABET council approves merger with CWA. *Daily Variety*, p. 1.

O'Steen, K. (1992b, October 7). NABET, IA merger is final. *Daily Variety*, p. 3.

O'Steen, K. (1992c, December 14). SAG and AFTRA again discussing merger. *Daily Variety*, p. 4.

O'Steen, K. (1994, February 7). State losing post-prod'n. *Daily Variety*, p. 12.

Papp, L. (1992, October 18). Unions at crossroads: Canadian organizations say survival lies in merging forces against hostile employers. *The Ottawa Citizen*, p. E6.

Prindle, D. F. (1988). *The politics of glamour: Ideology and democracy in the Screen Actors Guild*. Madison: University of Wisconsin Press.

Raskin, A. H. (1988, December 26). Review: The politics of glamour. *The New Leader*, p. 21.

Robb, D. (1992, October 1). Berman quip about Canada draws FIA ire. *The Hollywood Reporter*.

Robb, D. (1993, July 26). Mexico studio makes DiTolla wary on NAFTA. *The Hollywood Reporter*.

Robb, D. (1994, May 23). Unions trying to get merger talks on track. *The Hollywood Reporter*.

Ross, M. (1941). *Stars and strikes: Unionization of Hollywood*. New York: Columbia University Press.

Rowan, R. L., Pitterle, K. J., & Miscimarra, P. A. (1983). *Multinational union organizations in the white-collar, service, and communications industries*. Philadelphia, PA: The Wharton School.

Schwartz, N. L. (1982). *The Hollywood writers' wars*. New York: Knopf.

Span, P. (1992, December 29). N.Y. feels pinch of lost filmmaking. *Los Angeles Times*, p. F8.

Staiger, J. (1981). *The Hollywood node of production: The construction of divided labor in the film industry*. Unpublished doctoral dissertation, University of Wisconsin, Madison, WI.

Storper, M., & Christopherson, S. (1987). Flexible specialization and regional industrial agglomerations: The case of the U.S. motion picture industry. *Annals of the Association of American Geographers*, 77(1), pp. 104-117.

Stremfel, M. (1989). Decline of union power accelerates in Hollywood. *The Los Angeles Business Journal*, 11(8), 22.

Stumer, M. B. (1992, August). Show-Biz unions, guilds: A practitioner's guide; cutting-edge issues. *Entertainment Law & Finance*, p. 1.

Sumner, J. (1993, December 3) Houston gets the biggest bucks from film and TV production. *The Dallas Morning News*, p. 5C.

Time Warner. (1992). *Annual Report*.

Ulmer, J. (1993, November 19). Hollywood: NAFTA's tariff-ic. *The Hollywood Reporter*.

Universal Studios stagehands reject representation by IATSE, craft council. (1993). *Daily Labor Report, No. 68*, p. A-8.

Wasko, J. (1983). Trade unions and broadcasting: A case study of the National Association of Broadcast employees and technicians. In V. Mosco & J. Wasko (Eds.), *The critical communications review* (pp. 85-114). Norwood, NJ: Ablex.

Wasko, J. (1994a). Jurassic Park and the GATT: Hollywood and Europe-An update. In F. Corcoran & P. Preston (Eds.), *Communication and democracy in the New Europe* (pp. 157-171). Cresskill, NJ: Hampton Press.

Wasko, J. (1994b). *Hollywood in the information age: Beyond the silver screen*. Cambridge, MA: Polity Press.

Wasko, J. (1994c). "Hollywood Goes East." In K. Jakubowicz (Ed.), *The audiovisual landscape of Central and Eastern Europe* (pp. 75-89). Antwerp: Audiovisual Eureka.

Welkos, R. W. (1993, December 30). Craft unions, film producers reach new pact. *Los Angeles Times*, p. D4.

9

Beyond the Last Bastion: Industrial Restructuring and the Labor Force in the British Television Industry*

James Cornford and Kevin Robins

The British television industry has undergone more than a decade of rapid change, driven by the regulatory transformations brought about by the free-market Thatcher governments during the 1980s. Thatcher once famously described the television industry as "the last bastion of restrictive practices" (quoted in Davidson, 1992, p. 10). By way of general trade union legislation, ideological offensives, and regulatory and legal change to reshape and restructure the television industry, the Thatcher governments markedly reduced the power of organized labor within the industry.

---

*The authors acknowledge the support of the Economic and Social Research Council's Programme on Information and Communication Technologies (PICT). Thanks also to Andy Egan at BECTU and Jonathan Davis at London Economics for their assistance.

BRITISH BROADCASTING

What was the structure of the British television industry that the neoliberal Thatcher governments inherited? In the 1970s, British television was a highly stable system, based on a set of carefully constructed and rigorously policed compromises. At the heart of this structure was the division between the publicly funded British Broadcasting Corporation (BBC) and the commercially funded Independent Television (ITV) system.[1]

The BBC, with two television channels, was funded out of the proceeds of a license fee levied on all owners of television sets. The spread of television, and then the replacement of black and white televisions with color sets (which carried a higher license fee) meant that, until color television ownership reached saturation level in the early 1980s, the BBC's income was always rising. Virtually all the domestically produced programming shown on BBC television was produced in-house by full-time employees. The corporation was thus able to provide secure employment and a clear career path for creative and technical personnel. It also undertook a range of other functions for the industry as a whole, including much of the technical and professional training. The BBC's employees were mainly represented by the Broadcasting and Entertainment Trades Alliance (BETA, which also represented theatrical workers and others). Although the corporation was government owned, it was established under a Royal Charter with its own board of governors appointed by the responsible minister of the day and was consequently kept at "arms length" from direct political interference.

The ITV system comprised a network of 15 separate commercial companies of various sizes that collaborated to create a national network channel, each selling advertising airtime in its designated region. As with the BBC, virtually all the programming that was specifically produced for ITV was produced in house by one of the companies. These companies, each of which enjoyed a regional monopoly on advertising airtime sales, proved to be very profitable (one owner, Lord Thomson of Scottish Television, famously described commercial

---

[1]This division was mirrored in the structure of the unions that organized in the television industry. The BBC's technical workers were mainly represented by the Broadcasting and Entertainments Trade Alliance (BETA). The Association of Cinematograph and Allied Technical Trades (ACTT) was the main technical union in the ITV system and the film industry, but was not recognized by the BBC. At the end of the 1980s, the two unions merged to form the Broadcasting Entertainment Cinema and Theatre Union (BECTU). Other unions—for example, the Electricians (EEPTU), Actors (Equity), and Journalists (NUJ) unions—also organized within the television industry.

television as "a licence to print money"). The main union representing technical and creative grades in the ITV companies was the Association of Cinematograph and Allied Technical Trades (ACTT, which also organized within the film industry).

The ITV companies were particularly susceptible to organized trades union action. If the highly unionized television technicians "pulled the plug" on an ITV company, advertising would not be seen by viewers, and the dominant source of revenue for the company would consequently dry up immediately. (By contrast, the BBC would actually save money in the event of an industrial dispute as the license fee on television sets was collected whether there was a service or not.) Wage rates within ITV tended, as a result, to be consistently higher than in the BBC, as ITV management generally found it easier to buy off union demands with extra money and then use their monopoly power to simply pass the costs on to advertisers (who would, of course, pass them on to consumers). Although the ITV companies were private commercial undertakings, their licenses to broadcast were tightly regulated by a statutory body, once again appointed by the government—the Independent Broadcasting Authority (IBA).

Until the 1980s, the BBC/ITV duopoly was a closed, stable, and highly integrated system. Virtually all producers worked in either the BBC or the ITV companies. There was no competition between the two parts of the duopoly for revenue, in spite of stiff competition for viewers. Even within the ITV system, competition for advertising revenue was muted by the allocation of regional monopolies. Imported programs were shown on both channels, but they were limited by a "gentlemen's agreement" to around 18% of screen time. Although there were tensions in this cozy arrangement, each of the major interest groups in the television industry felt they were getting a reasonable deal: Management and ITV shareholders got a successful industry with, in the case of ITV, high and relatively stable profits; producers and workers got a degree of freedom to make a variety of types of well-funded programs as well as high wages (ITV) or job security (the BBC).

THE GOVERNMENT AND MANAGEMENT OFFENSIVE

The stable set of compromises that had sustained broadcasting in Britain until the 1980s met their nemesis in the Conservative governments led by Margaret Thatcher, which ruled throughout the 1980s. The brand of conservatism that Thatcher represented was constructed around the twin goals of a free market and a strong (but minimal) state. From this neoliberal perspective, the British government saw as its major challenge the encouragement of a market-led restructuring of British industry. Management was to be free to manage, which meant that it must be free from both union power and excessive government regulation. Competition was to be encouraged and supported with industries opened up to the harsh winds of international competition. Enterprise was to be promoted, both within large firms and through the creation of new small firms. New technology was to be embraced as an aid to competitiveness and as a means to undermine the power of organized labor. The law offered a further weapon that would be used to hamper and restrict the operations of the trades unions.

Communications workers—in both broadcasting and the press—came to be one of the prime targets of a combined government and management offensive, only toward the end of the 1980s. The strategy of the Thatcher governments was to pick off bodies of organized labor one by one. (Thatcher is reputed to have said that she did not mind having enemies, but she did not want them all at once.) During the first Thatcher administration (1979-1983), the power of the unions in the manufacturing industry was dramatically reduced by the recession of the early 1980s, which created high levels of unemployment and a dramatic shakeout of capacity in the main industrial centers of the North. The second administration (1983-1987) saw the year-long (1984-1985) miners' strike, which effectively destroyed the National Union of Mineworkers, the union the government feared most. It was not until after these famous victories that an offensive commenced against the workers in the communications sector.

British communications workers were seen as an especially important target by the government. The national newspaper industry, although overwhelmingly supportive of the Conservative Party at election time, was particularly strongly unionized, with a highly complex set of demarcations and a range of infamous "Spanish practices," which included almost total union control over the hiring of print workers and the paying of wage packets to nonexistent "ghost" employees (which were then shared among the real workers). The television industry was, if anything, even more of a challenge to the Conservative neoliberal ideal. In the context of this market ideology, the stable, closed BBC/ITV duopoly—with its complex bureaucracies, liberal public service ethos, careful limiting of competition, and its acceptance of the legitimate role of organized labor within the television industry— was clearly an anathema.

The government's determination to undermine the power of organized labor in the communications industries was significantly hampered by a range of factors. Unlike manufacturing and coal mining, neither newspapers nor television faced significant foreign competition, so there was little hope that the forces of competition would flow directly from that quarter. What is more, there were significant divisions within management cultures and perspectives within the communications industry: In the press these tended to be between "owners" and "editors"; in broadcasting a much sharper distinction could be drawn between "program-makers" and "accountants." The tradition of political and editorial independence in the British media meant that the balance of power between these different sections of management had tended to fetter the more commercial instincts of owners and accountants.

With products that had to come out every day if revenues, especially advertising, were not to dry up, management was forced to come to some sort of compromise with labor. Internally riven in this way, and with little chance of surviving determined industrial action from the trades unions, management had pragmatically accepted the unions as fact of life. The newspaper proprietors and ITV management had always found it easier to buy off action by the technical unions (print workers or television technicians) than to attempt to force through new working practices and new technology. The result was a system of labyrinthine agreements between management and the unions and a resistance to new technologies, which were associated with deskilling and labor shedding. In both broadcasting and the press, management had learned to live with union power (which, as long as it applied to all managements and proprietors equally, would not affect their relative competitive positions).

Such pragmatic acquiescence to union power was galling to all true Thatcherites. In seeking to undermine the power of the unions in the communications sector, the government had to stiffen the resolve of management. One source of more determined management was from "outsiders"—regional newspaper proprietor Eddy Shah and Australians, Rupert Murdoch and Bruce Gyngell. It was these relatively new interests that, with more or less tacit encouragement and assistance from the government, mounted a determined management offensive. At the same time, the government sought to shift the balance of power within the established core of the industry and to push the owners and accountants into the ascendancy over the more liberal editors and producers. This initiative took both direct and indirect forms. Direct intervention such as the appointment of businessmen to chair the Board of Governors of the BBC (Marmaduke Hussey) and the IBA (George Russell) was complemented by slowly increasing financial pressures on the BBC and the ITV and by new trade union legislation designed to strengthen the hand of management in any future dispute.

The first real attempt to break the power of workers in the communications industry occurred in the middle of the decade when a regional newspaper proprietor, Eddie Shah, determined to overcome the influence of the print unions by using the new trade union legislation that had been introduced by Thatcher in 1980 and 1982. Shah installed new color presses and came to an agreement with the Electricians Union that its members, and not the traditional printers, would operate them. The result was several months of picketing, sometimes violent, outside Shah's Warrington plant and then a long running legal case that saw the National Graphical Association (NGA), the main printers union, fined half a million pounds. The Electricians Union, and its leader Eric Hammond, became deeply unpopular within the trade union movement and was, for a time, expelled from the Trades Union Congress, the collective body of organized labor in Britain. Warrington was "probably the first time that an industrial dispute within the newspaper industry had cost the unions more than the owners" (Shawcross, 1992, p. 336).

The lessons of Warrington were not lost on the proprietors of the national press. The first to make a move was Rupert Murdoch. In 1985, Murdoch was persuaded by his executives to move the printing of his four main national titles—*The Sun, The News of the World, The Times,* and *The Sunday Times*—away from their traditional Fleet Street location to a new, especially prepared, site at Wapping in London's East End (a similar plant was prepared in Glasgow to print the Scottish and Northern editions). Under extreme secrecy, tried and tested technology that would allow direct input of copy was installed, and the new site was fortified against picketing and disruption. Once again, the Electricians Union was used to operate the presses, and workers were recruited from Southampton, far from London, in order to prevent the established print unions from realizing what was about to happen. A vast fleet of 800 lorries and vans was assembled to distribute the papers produced at Wapping, thereby avoiding the possibility of a sympathy strike by the railway workers who had formerly delivered them throughout Britain. At the start of 1986, after all the arrangements had been made, production of the four titles got underway at the Wapping compound. Although there were a number of high-profile resignations and refusals, most of the journalists on the four papers were enticed, or cajoled, into the move to Wapping by increases in their pay packets.

With workers in the Electricians Union now producing the four national papers at the Wapping plant, the print workers at the Fleet Street sites went out on strike. In retrospect, it can be seen that this move played directly into Murdoch's hands as, using the Thatcher government's trade union legislation, he was not only able to fire them all for breach of contract, but also to avoid payment of any redundancy money. In the ensuing six months, the notorious Wapping site became the scene of mass picketing and, on occasion, of horrific violence as the picketers attempted to impede journalists, electricians, and others from entering and leaving the plant and to prevent the lorries carrying the newspapers from getting out of the East End. The police, forewarned by Murdoch, fought to maintain access to and from the Wapping plant. Meanwhile, the unions were taken to the courts, found in contempt, and fined (e.g., SOGAT had all its assets impounded). By the middle of 1986, it was clear that Murdoch had won. Thatcher was reportedly ecstatic. Other newspaper proprietors were quick to follow Murdoch's lead.

With the print unions vanquished, there remained only one significant outpost of real union power in the communications industry: the television technicians, and above all the technicians in the ITV system. For Thatcher, speaking in 1987 at a special seminar for British television's senior management, the television technicians were "the last bastion of restrictive practices" (quoted in Davidson, 1992, p. 10). In making such a charge, she was implicitly calling for a response, both from her ministers and from the broadcasting industry managers assembled before her. She did not have long to wait. In November of that year, a dispute broke out at the ITV morning station—TV-am. The cause of the dispute, as with those in the newspaper industry, was the introduction of new technology—specifically the introduction of remote-controlled cameras—which was seen as threatening employment and skill levels within the company. The TV-am management, led by Australian Bruce Gyngell, locked out striking technical workers and, with the aid of imported programming from the United States, ran a scratch service for four months, with management taking over from the technicians. At the end of the dispute, the striking technicians were fired and replaced by a much smaller number of younger, cheaper, nonunion workers. Traditional demarcations were abolished, and the new technology was introduced. As a consequence, the company's wage bill was dramatically reduced. In commenting on TV-am's next annual report, the *Financial Times* observed that it had "achieved most saving by replacing its 172 technicians employed in jobs such as camera operation and vision control with 63 multi-skilled operators who mostly earn between £10,000 and £21,000" (quoted in Gapper, 1990).

The example of the TV-am lockout was certainly not lost on the management of the other ITV companies. However, the possibility of replicating the TV-am strategy of using imported programming and setting managers to operate cameras and other equipment could only have worked at the morning station with its wholly studio-based output. Other strategies would have to be pursued by the main ITV stations.

While TV-am opened up one avenue of attack on organized labor, the government was busy opening up another. In March 1988, the Secretaries of State for the Home Office and for the Department of Trade and Industry (the ministers responsible for the television industry and the film industry, respectively) referred the question of "restrictive labor practices" in the film and television industry to the Monopolies and Mergers Commission (MMC) under section 79 of the Fair Trading Act 1973 (the first use of this section of the Act against a group of trades unions). This legal approach was clearly designed to maximize the pressure on the unions to acquiesce in the introduction of new technology, the development of "multiskilling," and the consequent scrapping of established demarcations and rules about staffing levels.

When the 113-page MMC report finally arrived in 1989, the unions were substantially exonerated. Although the MMC had found that there had been "restrictive practices" in the past, it also noted that these practices had either already been abandoned or were in the process of withering. Overall, the Commission reported that its "conclusions concerning the public interest have been reached against the background of . . . fundamental changes in the industry. . . . These changes reflect an increasingly competitive environment to which employers and unions alike have committed themselves to respond flexibly" (Monopolies and Mergers Commission, 1989, p. 2). The MMC inquiry clearly had the ideological effect on the unions that the government had intended, placing them on the defensive.

THE RESTRUCTURING OF BRITISH BROADCASTING

What had happened to turn "the last bastion of restrictive practices" into an industry in which the unions had "committed themselves" to "respond flexibly" to industrial change? The management at the morning broadcaster, TV-am, had been able to ride out union action because of the large volume of paid-for programming and the studio-based nature of the service. Managements at the main production centers of the BBC and the ITV companies, which were far more reliant on their own technical employees to provide a service, had to tread much more carefully in their relations with the unions. Nevertheless, a major process of industrial restructuring within the television sector was gradually stacking the deck in their favor.

The management offensive against the television technicians coincided with, and was underpinned by, a significant structural change in the way British television organized its activities, involving the production of programming outside the formal structures of the BBC and the ITV companies and the growth of a new "independent" production sector. In the 1970s, there was only a small number of independent production companies supplying British broadcasters, mainly acting as vehicles for individual star presenters. By 1994, membership of the producers' trade association (PACT) reveals that there were over 1,200 independent production companies in Britain. It was the development of this sector of mainly small producers and facilities companies, and of a freelance labor market associated with them, that opened the first major breach in the organizational defenses of the BBC and the ITV companies.

The independent production sector was effectively created by the decision to establish, in 1982, the new fourth channel, Channel 4, as a "publisher-contractor" broadcaster (Lambert, 1982). This meant that, unlike the BBC and ITV companies, the new channel would have no production facilities or staff of its own. Instead, it would commission programs from external producers—"independents." Thereby, it was argued, the new channel would open television up to formerly excluded voices and help to maximize resources—both in terms of facilities and labor—that had developed around the film and corporate video industries. Most independent producers were lightweight organizations, with permanent employment of only three or four principle individuals and a rented office; all other requirements were sourced from outside— labor from the freelance market, facilities from independent facilities houses, and so on. There was the belief, in many quarters, that the independent sector could contribute both creativity and efficiency to the British television industry (on the independent sector, see Robins & Cornford, 1992).

So long as independent production was restricted to Channel 4, however, it acted as little more than a supplement to the existing production activities within the BBC and ITV companies. Channel 4 was carefully protected from the full rigors of the market by having no private shareholders (it was, and still is, in effect government-owned) and by the establishment of a funding system that gave the ITV companies responsibility for selling the airtime on Channel 4 within their own regions (and therefore a continued monopoly over television advertising in their regions). In 1987, however, the government decided that it would extend the Channel 4 principle, pointing to the success of Channel 4 in both creative and financial terms. Quotas were imposed for 25% of BBC and ITV production to be contracted out to independent companies by 1992. As the BBC and ITV dutifully sought to meet their quotas, the independent sector came to be increasingly influential across the entire British television industry.

The Conservative government's perspective on independent production was most clearly presented in its 1988 white paper, *Broadcasting in the '90s: Competition, Choice and Quality* (Home Office, 1988). In this document the government claimed that "independent producers constitute an important source of originality and talent which must be exploited [sic] and have brought new pressures for efficiency and flexibility in production procedures" (p. 44). With this in mind, the White Paper proposed that "there should be a greater separation between the various functions that make up broadcasting and have in the past been carried out by one organisation" (p. 44). As Colin Sparks (1989) has pointed out, in this passage "the key words are obviously not the praise for 'originality and talent' but the more familiar terms like 'exploit', 'efficiency' and 'flexibility'" (p. 35).

Independent producers, then, represented a mechanism that could be used to counter the "excessive degree of vertical integration" (Home Office, 1988, p. 6), which the government saw as generating rigidities in the British television industry. It was on this basis that the 1990 Broadcasting Act gave statutory force to the 25% independent production quota for all terrestrial broadcasters (BBC and ITV) to be attained within five years. For the government, support for independent producers was associated with breaking up the "cozy duopoly" and creating a new era of competition and expansion in broadcasting.

Although the broadcasters were at first unenthusiastic about independent production, a number of changes in the wider broadcasting environment began to win them round to an appreciation of the advantages that it might hold for them. For the ITV companies, this change of attitude was occasioned by the anticipated impact of the 1990 Broadcasting Act. The Act legislated that henceforth the ITV licenses should be auctioned to the highest bidder (subject to some assurances on the feasibility of the business plan and the quality of the programs offered).[2] This idea of auctioning franchises had been pushed through by the Treasury, against almost universal opposition from the television industry, in order to increase the amount of money that was extracted from the ITV monopoly system (the ploy was so successful that the

---

[2]The ITV companies have been periodically required to reapply for their licenses in an open competition. In previous competitions for licenses, the procedure was for the regulatory body to hold a "beauty contest" in which each company would promise to produce specific programs. The regulator would then award the license to the company that was regarded as offering the best "quality" service (and therefore, generally, would spend more money on program creation). By moving to a cash auction for licenses, the government was implicitly encouraging companies to reduce their planned expenditure on program creation so as to make a maximum cash bid.

Treasury got an estimated additional £100 million from ITV in the first full year of operation of the new Act). With so much more money being siphoned out of the television industry by the Treasury, the ITV companies became desperate to economize in every area of expenditure—and that meant producing programs, among other activities.

At the same time, the new Act broke the established principle of no direct competition for advertising revenue among terrestrial broadcasters by separating Channel 4 from the ITV system and allowing it to sell its own airtime. The ITV companies thus faced significant direct competition for advertising revenue for the first time. What is more, after a very slow gestation, the cable and satellite industries were beginning to develop much more rapidly, threatening a further source of competition for advertising revenue when their aggregate audience reached an appropriate size. From being protected regional monopolies, the ITV companies now faced the prospect of remorselessly increasing competition for their main source of revenue, the sale of television airtime. All of this coincided with a deep and sustained economic recession, which led to a significant weakening in demand for television advertising airtime. In this advertising famine, and with ever stronger competition looming on the horizon, cost cutting became even more imperative. The obvious target for "efficiency" measures were the professional and technical workers, who represented the major cost for the ITV companies.

For the BBC, although still protected from the prospect of competition for revenue by its monopoly on the license fee, things were not much better. The Conservative government consistently regarded the BBC as a rigid and unentrepreneurial monolith that was wasting an excessive amount of the license fee on bureaucracy. The government was also highly aware of the unpopularity of the license fee. Regular increases in the license fee, which were designed to take account of inflation, were frozen, and future increases were pegged to a rate below the retail price index. After years of rising real income, the corporation faced a major squeeze on its budgets for the first time. Here, too, constantly increased "efficiency" and cuts in staff numbers became the main priority. Again, the labor force was bound to feel the consequences.

For both the BBC and ITV companies, then, cost saving became a number one priority and, with this in mind, both organizations came to see the virtues of independent production. In this climate, the balance of power between accountants and producers shifted decisively in favor of the former. The existence of independent production enabled television managements to restructure their own activities in order to attain higher levels of efficiency through the introduction of accurate cost accounting for programs, through divisionalization and the introduction of internal markets, through the introduction of new technology, and through an increasing reliance on freelancers. The growing use of independent producers to supply programs had, for the first time, given the accountants a yardstick against which to measure the "efficiency" of their own internal production units. If an independent producer could do it cheaper, then the clear financial imperative was to commission the program from the independent producer. With the potential for almost any programs (except the sensitive and high-status news programs) to be "put out" to independent producers, remaining full-time workers were inclined to be much more cooperative about wages, demarcations, and the introduction of new technology. Furthermore, as demand for independent production grew, supply increased faster, bringing about intensified competition within the sector and driving down the prices that many independent producers could command. Because the accountants regarded the independent sector as the yardstick of what it cost to produce programs, the pressure for even greater efficiency from internal units continued to mount.

As more production was transferred from the BBC and ITV companies into the independent sector, the levels of permanent employment required by the broadcasters were consequently reduced. Some of these reductions were the result of moving ancillary services such as cleaning, security, and catering from in-house operations to outside contractors. For the most part, however, it has been the producers and technicians themselves who have been made redundant, either to be replaced by freelancers or, in many cases, simply rehired on short-term contracts. The BBC and ITV companies have thus begun, in the name of efficiency, to develop the same relationship with skilled labor that the independent production companies pioneered, hiring workers for as long as required and no longer. Even when permanent employees have been retained, in-house production and facilities units are being increasingly forced to act like independents: Split up into cost centers, they have been expected to compete for work in both internal and external markets. Furthermore, as in-house staff numbers have fallen, management has increasingly insisted on the scrapping of demarcations between different tasks in the name of multiskilling to ensure that those workers still on full-time contracts are utilized to the full.

At the BBC this process has been most clearly pursued through the introduction of an "internal market" under the label of Producer Choice. Producer Choice originated with a report commissioned by senior management from the corporation's then head of Finance, Ian Phillips. The Phillips Report was intended to assess ways in which the

BBC could increase efficiency and therefore live within an income that was declining in real terms. Phillips' recommendations centered around the creation of an internal market within the organization, in which management would allocate cash budgets rather than goods and services. Producers would be able to source their requirements either from within the corporation or from outside. In pursuit of this policy, more than 400 separate "business units" were established. The point was to transmit market pressures for efficiency into the heart of the BBC. Clearly, this policy is having an effect. Even using the BBC's own, somewhat misleading, figures, it is apparent that, by 1992/1993, the average employment level in BBC television fell to just 83% of its 1986/1987 value (see Table 9.1).

The ITV companies have been even more drastic in their shedding of full-time workers. Table 9.2 gives some indication of the increasing pace of the "slimming down" process that has taken place. By 1992/1993, employment levels had slumped to just 60% of those of 1987/1988. Even so, the data seriously underestimate the real impact by failing to disentangle true permanent staff from "average staff numbers," which include those on freelance contracts. Hence, to take just one particularly dramatic example, Tyne Tees Television's average staff employment moved from a high point of 700 employees in

Table 9.1. Average Employment in BBC Television, 1983-1993.

Year	BBC Television Employment	Index (Percentage of 1986/7 Figure)
1983/4	17,138	94%
1984/5	17,679	97%
1985/6	17,992	99%
1986/7	18,243	100%
1987/8	17,908	98%
1988/9	17,272`	94%
1989/90	16,940	93%
1990/1	16,504	90%
1991/2	16,232	89%
1992/3	15,092	83%

Note that these figures are not strictly comparable. Reporting conventions vary year by year. For example, in 1991, some 1,070 part-time employees were included in the total as 562 full-time equivalents. The table should thus be read with caution.

Source: Data compiled from BBC Handbooks and Annual Report and Accounts, various years.

Table 9.2. Average Employment in ITV Companies 1987/1988–1991/1992.

Company	1987/8	1988/9	1989/90	1990/1	1991/2	1992/3
Anglia	876	876	875	797	738	663
Border	239	256	231	174	144	129
Central	2,043	1,977	1,802	1,521	1,203	942
Grampian	364	356	289	222	190	188
Granada	1,533	1,456	1,345	1,345	1,451	1,363
HTV	1,240	1,280	1, 360	1,392	1,226	825
ITN	978	1,116	1, 125	1, 219	1,267	987
LWT	1, 677	1, 631	1,351	1, 287	1,078	1,039
Scottish	804	806	743	781	787	650
TSW	380	361	315	315	290	190
TVS	1,097	1,053	1,076	903	720	431
Thames[a]	2,712	2,654	2,470	2,546	2,113	1,321
TV-am	469	370	416	423	427	256
Tyne Tees	721	696	661	666	476	b
Ulster	303	303	299	293	299	252
Yorkshire	1,702	1,623	1,471	1,350	1,310	1,067
Total	17,138	16,814	15,829	15,243	13,719	10,303
Index (percentage) of 1987/8 figure	100%	98%	92%	89%	80%	60%

Note: Conventions for calculating staff, year-ends, and other factors vary. This table should therefore be read with caution.

Source: Data compiled from Broadcast, various years

[a]Includes Thames subsidiaries.
[b]Included with Yorkshire

1987/1988 to just 198 by April 1993, following the takeover of the company by the neighboring Yorkshire Television (Courtice, 1993). With the replacement in 1993 of three integrated ITV producer broadcasters (Thames, TVS, and TSW) by commissioner-broadcasters (Carlton, Meridian, and Westcountry), the number of permanent jobs within the ITV system is set to fall even further. As further mergers and takeovers loom, and as the merged companies remove duplicate functions, further shedding of permanent jobs will surely result.

What of new sources of employment in the expanding cable and satellite television sectors? At the beginning of the 1980s, British viewers had a choice of three terrestrial television channels, none of which broadcast a morning service and all of which were dark during much of the night. By the early 1990s, this situation was transformed with four terrestrial channels—three of which were offering a morning service and two of which were broadcasting for 24 hours a day—and a further 20 or more channels were available from satellite or cable. However, this massive expansion in volume of broadcast television was not being translated into an equivalent increase in demand for new production. Instead, the new satellite- and cable-delivered channels such as BSB and Sky broadcast repeats and cheap studio-based programs. As Peter Chippendale and Suzanne Franks (1991) explain:

> From the mid 1980s, television companies were busy trying NOT to make programmes, and although BSB may have seemed like a television company to the outside world, it was actually a clearing house and publisher of programmes that were either specially commissioned or bought in the old "acquired" way. Sky had begun going down a similar route, gathering large amounts of acquired programming, the joy of which was they could be very cheap indeed. (p. 111)

Even when new programming was commissioned, it was produced on a shoestring budget. For example, in the late 1980s, the average cost of network television for the BBC or ITV was over £82,000 per hour; Murdoch's Sky, by comparison, was claiming that it could manage a schedule on just £6,000 per hour (Chippendale & Franks, 1991). With the "merger" of Sky and BSB in November 1990, even less newly produced programming was required. Of course, the satellite channels were predominantly nonunion.

The net effect of this overall process of restructuring has been to casualize the workforce, or a large part of it. The freelance market, in which contracts can be measured by hours or days, has expanded rapidly. According to a recent union study, the majority of television workers are, for the first time, freelancers, and that proportion is still growing:

> In 1959, two-thirds of the 8,000 membership [of the ACTT] were mainly *employees* in the film laboratories and television companies, with the remaining one-third mainly freelance. By 1992, the membership can be expected to reach 33,000, with two-thirds mainly *freelance* working in film and television. (Grainger, 1991, p. 35)

The decline in the number of permanent posts has not been fully met by the increasing volume of freelance work available. This reversal of the ratios of freelancers to permanent employees has therefore had a number of implications. First, it has led to a dramatic increase in the level of unemployment and underemployment in the sector. A survey carried out by the ACTT in the mid-1980s indicated that longer term underemployment in the freelance sector (with members being unemployed for more than six months in the preceding year) was growing rapidly (see Table 9.3). By the early 1990s, survey results and informed estimates suggest that average utilization in the freelance television sector had reached about 50% (i.e., that the *average* freelance worker is only working six months in the year).

This casualization of the labor market has undermined the capacity of unions to provide a structure for the labor market. In the past, the union membership ticket acted as a form of accreditation, indicating that the holder had the professional skills and experience to undertake particular tasks. Because the union "ticket" had this status, union membership provided a kind of filter that restricted the size of the labor market. This system ensured more or less full-time employment, even for freelancers, because union membership would only be expanded if there was adequate extra demand from the employers. Under the new conditions, the unions have found it difficult to recruit and retain members from the greatly expanded number of freelancers. At the same time, some independent producers have proved willing to use nonunion labor and have avoided union agreements (by failing to join the independents' trade association, PACT). According to Roger Bolton, general secretary of the Broadcasting, Entertainment and Cinema Trades Union, "we need more freelancers and we need to persuade production companies to become members of PACT and to agree to industry rates and conditions" (quoted in Baker, 1993, p. 15). Without such changes, the union's ability to control entry and maintain employment standards to the labor market is severely impaired.

The most important implication of the growing proportion of freelancers in the industry has been to undermine the capacity of the industry to maintain its training and skill base. Freelancers face a particular problem: When they are working they will have the money but not the time for training; when they are between contracts they will

Table 9.3. ACTT Freelance Membership: Time Unemployed Over the Preceding Year.

	1983-1984	1986-1987
Unemployed over 2 weeks	56	42
Unemployed over 6 months	9	15

Source: ACTT Survey of Freelance Members' Earnings, Work and Unemployment 1988.

have the time but not the money (see Vaarlam, Pearson, Leighton, & Blum, 1989). In effect, the British television industry has been running down its most valuable asset—the skills of its workers. As Sparks (1989) pointed out, "the pool of skilled labour did not come from nowhere; in the main it was trained by the BBC and honed its skills in the ITV companies" (p. 36). Although the industry has recognized that this problem exists, and although there has been the development of a "freelance training fund" sponsored by the BBC, ITV, Channel 4, and PACT, and a government-backed training and accreditation system of National Vocational Qualifications, these actions appear a woefully inadequate replacement for the old, union-backed system. Hence, although the television industry's skill base is eroded, "there is no sign of the state being prepared to invest in a training programme to compensate . . . as the pool dries up" (p. 37). At the same time, with no regulation of numbers entering the industry, and with the glamour of television encouraging a proliferation of spurious "media training courses" providing a steady stream of recruits, the laws of supply and demand are forcing down wage rates.

BRITISH COMMUNICATION WORKERS IN A GLOBAL INDUSTRY

The Conservative government's main motivation for breaking up the British television industry's stable duopoly was the desire to create an efficient industrial sector capable of competing in the increasingly open world television markets. It was felt that Britain could become the production center for an emerging "European audiovisual space"; it could become the "Hollywood of Europe." The reduction of wage levels and the promotion of labor flexibility through the destruction of union power within the British television industry was understood as a precondition of success for such a strategy:

> The argument runs that British companies have linguistic advantage in international competition in that they operate in the dominant international language and that, if wage costs can be forced down below international levels, there exists a large pool of highly skilled labour to produce exportable programmes which could find a market even in the US. (Sparks, 1989, p. 36)

This simplistic strategy presumed that all that was holding back the British television industry from European, or perhaps even world, dominance were over-mighty unions.

In preparing the British television industry for such an export-oriented strategy, independent production companies increasingly came to be used as agents of a market-oriented rationalization designed to reduce union power and cut wage costs. Their main function shifted "away from making new kinds of programmes towards making the same kinds of programmes more cheaply" (Davis, 1991, p. 76). Not only did the independents create a new source of cheap programs, they were also vital (if not wholly witting) instruments in reducing the volume of employment and/or wage levels in the remaining in-house production units through the weakening of union power. Sparks (1989) pointed out, as far back as 1989, that "the Government, certainly, is very pleased with the role of the independents in breaking the unions" (p. 35).

The results of this export-oriented strategy can only be judged a failure (see Figure 9.1). For the first time, the British balance of trade in television programs has dipped into the red. Far from promoting the

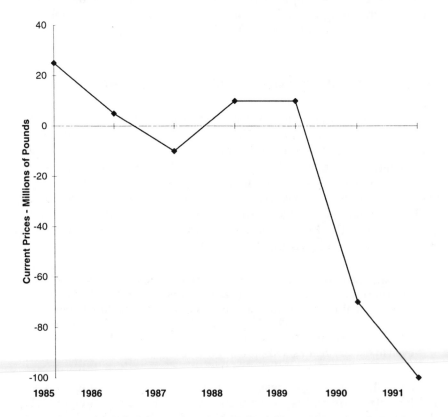

Figure 9.1. British balance of trade in television programs 1985-1991 (Current prices)

development of Britain as the powerhouse of European television production, the introduction of greater competition seems to have had the opposite effect. Strengthened competition between the BBC, ITV, Channel 4, and the new satellite- and cable-based broadcasters has driven up the costs of imported programs and rebroadcasting rights for sporting events and other programs. With more money required to purchase these elements in the schedule, less money has been available to produce or commission specially made domestic programming. Furthermore, greater competition for domestic viewers has compelled the channel controllers and commissioning editors at the BBC, ITV, and Channel 4 to concentrate on wooing the domestic audience with programs that have a particular national appeal. Hence, even when significant revenues are being put into domestically produced programs, those programs seem to have little transnational appeal and therefore little scope for export. At the same time, the extremely lax regulations governing the programming broadcast by "nondomestic" satellite channels, in particular the Murdoch-controlled BSkyB channels, means that satellite broadcasters can get away with showing a mix of low-cost, domestically produced studio programs with no export potential or imported films, dramas, and soaps.

There is, then, little evidence that lower wages and increased labor flexibility have done anything to benefit the competitive position of the British television industry in the new global audiovisual markets. There is some evidence that some U.S. studios have decided to shift a small part of their production activities to the U.K. in order to take advantage of lower wage rates. However, the volume of employment created by such relocation of U.S. production activity is small; there is always the threat that what little production that is attracted will leave as rapidly as it came, and there appears to be little substantial benefit to indigenous producers from foreign production attracted in this way. Although the volume of U.S. activity within Britain is likely to remain small, and will always represent a fragile source of employment, it could have its most significant effect on U.S. television workers. The major U.S. producers need only be able to *threaten* to shift production to Britain in order to bring pressure on U.S. communications workers to "moderate" wage demands and acquiesce to greater "labour flexibility." The real beneficiaries of the Thatcher government's deregulatory actions are thus the mainly U.S.-based, media multinationals, not the relatively small domestic ITV companies or the even smaller independent production companies.

Without the advantages of access to finance and the sheer distribution muscle that the major U.S. producers enjoy, cheap and flexible workers convey relatively little international advantage to

British television companies. In the end, by seeking to establish Britain as a low-cost, rather than high-quality, production base, the Conservative government increasingly pushed the British television industry into competition with Brazil, India, Mexico and other well-developed production centers (which, in at least some cases, also enjoy the supposed advantages of English-speaking technicians).

Even at the more restricted scale of the European Union, the benefits of the British government's "least-cost producer" strategy were more apparent than real. The European Commission's audiovisual policy stipulates that at least half of the output of European broadcasters must be made in Europe. This protectionist measure, having survived the recent, hard-fought Uruguay round of the GATT negotiations, may do something to encourage some forms of production in at least some European countries and to attract U.S. and other non-European producers to set up shop in Europe. However, even in this limited and protected European arena, Britain will face significant competition for the position of "least-cost producer." The Republic of Ireland, with the same linguistic advantages as Britain, is putting up a strong challenge to take up the role of offshore production base for the U.S. media multinationals forced by the imposition of production quotas to operate in Europe (see Bell & Meehan, 1989).

In encouraging the strategy of becoming the "least-cost European producer," the Conservative government closed off alternative strategies to positioning the U.K. within the global television industry. British television's traditional strengths—the production of a wide range of programming with high production values, based on skilled workers operating in a creative environment—gave the industry an opportunity to compete in the global audiovisual market, not on costs, but on quality. The attack on the unions undermined those traditional strengths and therefore damaged the capacity of the British television industry to compete on the basis of those traditional strengths. The increasing levels of competition introduced into the industry bred a "short termism," a concern for immediate efficiency and profits at the expense of the longer term future of the industry. Without adequate investment in the technical skills and creative capacities of the industry's workers, and without the "space" in which to develop and hone ideas before moving into production, the British television industry was in danger of losing its ability to sustain a position as a center of high-quality production.

The Thatcher-driven media policy has proven to be disastrous. The British television industry has been pulled out of the one race in which it had a reasonable chance of winning—the race to become a high-quality production center for the European and world markets.

Instead, it was pushed into a competition for the role of "least-cost" English-language production center, a race in which it had little chance of even finishing among the runners-up. British audiovisual policy has now gone so far down this path that the task of reorienting the whole direction of British television policy toward a strategy based on quality, not cost, will be a long and difficult one. Nevertheless, it is this task that British communications workers, and all those concerned with the future of the industry, will have to confront.

REFERENCES

Baker, M. (1993, October 22). BECTU's new broom. *Broadcast*, p. 15.

Bell, D., & Meehan, N. (1989). Cable, satellite and the emergence of private TV in Ireland: From public service to managed monopoly. *Media, Culture and Society, 11*, 89-114.

British Film Institute. (1993). *Film and television handbook 1994*. London: Author.

Chippendale, P., & Franks, S. (1991). *Dished! The rise and fall of British satellite broadcasting*. London: Simon and Schuster.

Courtice, G. (1993, May). "Tyne Tees" bosses and the pot of gold. *Stage, Screen and Radio*, pp. 18-19.

Davidson, A. (1992). *Under the hammer: Greed and glory inside the television business*. London: Mandarin.

Davis, J. (1991). *TV UK: A special report*. Peterborough, UK: Knowledge Research

Gapper, J. (1990, August 17). TV-am reduces wage bill by £6m. *Financial Times*.

Grainger, B. (1991, March). *The casualisation of the membership of the Association of Cinematograph, Television and Allied Technicians* (Policy Paper, No. 2). London: Centre for Communication and Information Studies.

Home Office. (1988). *Broadcasting in the '90s: Competition choice and quality* (Cm. 517). London: Her Majesty's Stationary Office.

Lambert, S. (1982). *Channel 4: Television with a difference*. London: British Film Institute.

Monopolies and Mergers Commission. (1989). *Labour practices in TV and film making* (Cm. 666). London: Her Majesty's Stationary Office.

Robins, K., & Cornford, J. (1992). What is "flexible" about independent producers? *Screen, 33*(2), 190-200.

Shawcross, W. (1992). *Murdoch*. London: Pan.

Sparks, C. (1989). The impact of technological and political change on the labour force in British television. *Screen, 30*(1&2), 24-38.

Varlaam, C., Pearson, R., Leighton, P., & Blum, S. (1989). *Skill search: Television, film and video industry employment patterns and training needs* (2 vols., IMS Reports Nos. 171, 186). Brighton, UK: Institute for Manpower Studies.

10

Hollywood North:
Film and TV Production
in Canada*

Manjunath Pendakur

With the restructuring of the world economy, the concentration, consolidation, and internationalization of Hollywood is rapidly transpiring. To cut the ever-escalating costs of production, Hollywood firms seek to export certain production processes, in some cases move entire production plants to areas of the world where labor costs may be lower. Labor under these conditions has to cope with the demands and vagaries of international capital and markets. Domestic cultural policies in countries such as Canada, previously enacted to preserve some elements of indigenous production and distribution, may be drastically affected. The case study presented in this chapter presents one such instance of how labor copes with the changing conditions in the film industry.

*This research was funded by a grant from the Canadian Embassy in Washington, DC, under its Faculty Enrichment Program.

213

Between 1920 and 1930, the U.S. film industry went through a period of intense consolidation when five vertically integrated firms dominated production, distribution and exhibition of motion pictures (Huettig, 1944). The structure that emerged has been resilient to the monumental changes in the national economy, the Great Depression, the two World Wars, such technological innovations as television, and even challenges from organized labor. The structure also extended into Canada and, despite periodic challenges, has remained intact there as well (Pendakur, 1990).

The heads of the leading vertically integrated companies dealt with labor's demands cunningly and brutally, often resorting to criminal methods. The major companies responded to the demands of labor by organizing a company union, what Ceplair and Englund (1979) called "management's most dog-eared trumps" (p. 18). The Academy of Motion Picture Arts and Sciences, which is respectably associated with the Oscar awards, was created by Louis B. Mayer of MGM in 1927 to contain labor militancy by bringing together producers, directors, writers, actors, and technicians. Apparently for five years, it successfully forestalled serious Hollywood labor organization.

The major firms also signed the Studio Basic Agreement in 1926 with the five largest unions—International Alliance of Theatrical and Stage Employees (IATSE or IA), United Brotherhood of Carpenters and Joiners of America, International Brotherhood of Electrical Workers, American Federation of Musicians, and International Brotherhood of Painters, Decorators, and Paperhangers of America. It was found that the Studio Basic Agreement (SBA) "served the executives well, for it maintained an open shop, centralized labor negotiations, and divided Hollywood unions into two ranks—those signatory to the SBA, and those excluded from it" (Hartsough, 1989, p. 52). If the objective was to keep a reserve army of labor, these policies succeeded because open shop meant that the production firm could hire nonunion labor when it suited its purpose, and that workers who were not members of these "recognized" unions need not be paid union wages. Although unions were waging their own jurisdictional and ideological battles, radical elements in these unions were isolated and faced summary dismissal and denial of jobs in the industry.[1]

The majors colluded in their attempts to control labor. For instance, when the Screen Writers Guild (SWG) was founded in 1933,

---

[1]The battles between AFL- and CIO-affiliated unions in the 1920s and 1930s, the internationals and the locals, democracy versus centralism on the shop floor, rank-and-file workers and their leadership all materialized in the development of labor relations in Hollywood. For an incisive analysis of those issues, see Ceplair (1989).

the majors refused to recognize it until 1938. The Supreme Court's ratification of the constitutionality of the National Labor Relations Act eventually forced their hand. Although writers got their first contract with the majors in 1941, they did not succeed in getting control over their own creative work as opposed to simply selling it as "product" to the producers. This meant not only that copyright belonged to the production firm but the power to recut or alter (or colorize as is done these days) the picture resided not with those who wrote or directed the film. The majors revived the IATSE in 1935 and brought the union under their control by bribing union leadership and getting "protection" from the Chicago syndicate (Hartsough, 1989). In the late 1940s, the most radical workers and their liberal counterparts in the SWG and other unions in Hollywood faced brutal repression with Congressional hearings on "communist subversion," ruining many lives in the purge that followed (Ceplair & Englund, 1979). McCarthyism effectively ended the potential political transformation of the industry by organized labor.

With little obstruction, internationalization of Hollywood has proceeded apace. Since the late 1980s, the trade press in the United States has extensively covered the story of U.S. "runaway" film and television production to Canada. Major motion picture studios and television networks have produced TV series as well as feature films in various provinces. CBS's miniseries, "I'll Take Manhattan," ABC's "Amerika," and Steve Martin's comedy "Roxanne" were all shot in Canada and were widely seen in both countries. One magazine account states that in 1987, a total of 28 productions were made in Canada by U.S. companies, of which five were feature films (Evans, 1988).

The locus of U.S. investment in Canada appears to be Toronto, Montreal, and Vancouver—the three principal centers of film production. In dollar terms, staggering investments have been made. Data sources conflict on the amount of U.S. investment. A Canadian magazine has, however, reported that total investment by U.S. companies in Canada for 1987 was C$282.9 million, of which B.C. received C$152.2 million (Evans, 1988).[2]

That is an unprecedented level of film activity by foreign companies in the history of Canada's motion picture industry. In the 1930s, there was a similar flurry of activity by U.S. film companies in response to the 1927 British Film Act. That law, concerned primarily with the growing dominance of the U.S. majors in the U.K., imposed a

---

[2]The Canadian dollar has steadily declined in relation to U.S. currency in the last two decades. In the 1980s, the difference was about 22 cents, whereas in the 1990s it has been about 35 cents. Canadian dollar figures are not converted into U.S. currency because of this differential and the possibility of creating the misimpression that the investment is not significant in terms of size.

screen quota. The U.S. corporations, however, found a loophole in the law, as films produced in the dominions qualified as "national" productions for the purposes of the quota. These films, popularly called the "quota quickies," were a temporary phenomenon. The British Parliament eventually plugged the loophole, thereby ending all such productions in Canada entirely (Pendakur, 1990). The current flood of U.S. investment in Canada's film industry appears to be radically different.

This chapter examines the differences in the nature of investment, external and internal reasons for the growth of production, and implications for Canada's cultural policy related to film/TV production. It focuses on the changing labor conditions, "flexible" policies toward transnational capital and working conditions, and what labor sees as strategies for survival in the changing global economy. By examining empirical data and interviews with policymakers in both government and labor unions, this chapter attempts to shed light on short- and long-term implications of these changes in relations between capital and labor in this important industry. The expansion of Hollywood production companies into B.C. and the jurisdictional disputes between certain unions provide useful case material in discussing the opportunities and conditions of labor in this industry.

The data were gathered from a combination of government, industry, and trade union sources. Film Commissions in Alberta and British Columbia and the Alberta Motion Picture Industry Association were good sources for information. Interviews with key actors in government, industry, and labor sectors of the film industry also helped fill in the gaps in printed sources and gave a human dimension to the problems and issues with which this study is concerned.

THE CONTEXT

It is useful to first take a brief look at Canada as a film and TV producing nation and its relationship to Hollywood. Figure 10.1 provides a snapshot of the Canadian film and television market in the context of the national economy in 1990. Until recently, Canada held the dubious distinction of being the number one market for film, television, music, and other entertainment from the United States. The Hollywood majors, which are the principal vertically integrated firms in production, distribution, and retail of these entertainment products, have successfully generated more revenue from other countries in the 1990s, as Canada slipped to Hollywood's fifth largest market. A total of 78.8 million movie tickets were sold in the country during 1990. Canada's theatrical box office gross for 1990 was estimated to be C$405 million, of which an average of 80% went to the U.S. majors, who are members of the Motion Picture Export Association of America.[3] The MPEAA

Population: 26.3 mil

GNP: C$629.4

Per capita income: C$23,987

Unemployment rate: 8.4% (Sept. '90)

Inflation rate: 4.2%

Savings rate: 10.8%

Fed deficit: C$30 bil

Trade deficit: C$1.7 bil (June '90)

Key population centers:
Toronto, Montreal, Vancouver

Gross Film B.O.: est C$405 mil

U.S. share of b.o.: 92-93% of English market; 70% of French market: national av. 80%

Total admissions (1989): 78.8 mil

Av. theater adm. price: C$7
in key first runs; C$7.50 in B.C.

Total households: 9.5 mil

TV in homes: 96.2%; 36.2% of homes in 1989 had two or more TV sets; no. of TV stations: 135

Pay out for imported TV programs: est. C$320 mil for US shows; VCR homes: 66% in 1989

Homes video rentals and sales: est C$1.278 bil; 92-93% for U.S. products

CDs, tapes and record sales: est. C$700 mil; 90% for U.S. products

Figure 10.1. Canada at a glance, 1990

---

[3]The MPEAA is a trade association, the membership of which is restricted to only some companies. In 1990, its members included Buena Vista International, Carolco Services, Columbia Pictures Industries, MGM/UA Communications, Orion Pictures International, Paramount Pictures Corp., 20th Century-Fox International, Universal International Films, and Warner Bros. International (see *The Hollywood Reporter*, 1990). With the acquisitions of New Line Cinema, a distributor, and Castle Rock Entertainment, a producer of theatrical films, Turner Broadcasting became a significant player in Hollywood. It was invited to join the trade association in 1995. MPEAA functions like a cartel abroad and is incorporated as a Webb-Pomerene association under U.S. law. The Webb-Pomerene Export Trade Act, passed in 1918, exempts such U.S. cartels from certain provisions of the antitrust laws of the United States.

member companies controlled a bigger share, at least 92% to 93%, in the English Canadian market for theatrical films.

Canada's theatrical exhibition market is dominated by two large chains with structural ties to the Hollywood majors. Famous Players (506 screens) is owned by Paramount Communications, and Cineplex-Odeon (563 screens) is a property of MCA-Universal.[4] Both chains have historically functioned as extensions of the U.S. market in Canada by way of ownership and certain mutually beneficial monopolistic trade practices employed by the U.S. majors.[5]

The television market in the country consists of 9.5 million households, of which 96.2% had one television set and 36.2% two or more sets. That market is serviced by the Canadian Broadcasting Corporation (CBC), a public service broadcaster; CTV and other private networks; and several cable operators, all privately held. Despite the Canadian content regulations, the TV market, at least during prime time, remains dominated by U.S. imports, and the revenue flow out of Canada is a considerable C$320 million for the imported U.S. shows.

Sixty-six percent of the homes had video cassette recorders, and Canada clearly appears to be the prime market for video rental and sales from the United States. An estimated C$1.278 billion was spent in video rental and sales in the country of which 92% to 93% was spent on imported materials from the United States. Canadians shelled out another C$700 million on recorded music in the form of compact disks, tapes, and records, 90% of that amount was in the hands of the U.S. majors.

Table 10.1 compares the three principal film and television production centers in Canada in 1990. Toronto has the largest number of movie screens and is considered the seventh largest market in North America for movies in terms of revenue potential by the U.S. majors. Vancouver, with a smaller population of 1.5 million, has almost twice the number of studios and nearly as many production facilities as Montreal. The reasons for that development is discussed later in the chapter. What is noteworthy, however, is that people of Quebec have historically supported French-language films, those made by both

---

[4]MCA-Universal was bought by Matshushita in Japan in the late 1980s and changed hands in 1995, when Seagram purchased 80% interest in that company. Nationality of ownership is often confused with national control and national interest, often an erroneous assumption. In this case, Canadian ownership of Seagram is not going to change how Cineplex-Odeon functions because its profitability is closely ties with the products from its principal suppliers, the Hollywood majors.

[5]For an analysis of these structural relationships and how they systematically produce underdevelopment in the unintegrated, Canadian-owned enterprises in production, distribution, and exhibition, see Pendakur (1990).

Table 10.1. Comparison of Three Principal Film/TV Production
Centers, 1990.

	Vancouver	Toronto	Montreal
Population (millions)	1.5	3.5	2.9
Studios	7	12	4
Production Facilities	20	23	22
Movie Theatres (Screens)	36 (97)	53 (239)	25 (99)
Legit theaters	18	25	39
TV Stations	3	13	4
Radio Stations	16	21	43

Source: In Adilman, 1990, p. 53. © 1990 *Variety*. Reprinted with permission.

French nationals and by Quebec artists. Quebec films, for instance, took
in a 20% share of the province's theatrical box office in 1989. The record
for English-Canadian films is strikingly different, with only 0.8% of the
box office share going to them (Adilman, 1990). Such success does not
necessarily translate into a lot of local production in Quebec. For
instance, in 1990, 36 features were in production in English Canada,
whereas only 10 were under production in Quebec (Adilman, 1990).
Quebec's generous tax shelter policy and a law that prohibits English-
Canadian and U.S. independent distributors from distributing directly
in Quebec have made it possible for some degree of local production in
the province.

INFLOW OF U.S. INVESTMENT

Accurate and complete data on the inflow of investment from the
United States and other countries to Canada are not available. It is
estimated that motion picture and television production adds C$1
billion to the Toronto economy, in which approximately 20,000 people
are employed in film-related activity. A Toronto study reportedly
showed that employment in this industry grew by 68% between 1983
and 1990 ("Movie makers," 1993).

In 1978, the total business conducted in B.C. as a result of motion picture and television production amounted to C$12.5 million (see Table 10.2). By 1991, the film industry poured C$176 million into the B.C. economy, and if we applied the indirect spending multiplier of three (the one most often used), the net expenditure totaled C$528 million.[6] Domestic nontheatrical productions (commercials and corporate videos) contributed another C$45 million. Employment of local talent hired on productions has grown from 40% in 1978 to 97% in 1991 (*Hot Property*, n.d.). The film/TV investments in B.C. contributed an estimated C$750 million to that province's economy in 1992 (Casselton, 1993).

Table 10.2. Productions in B.C. By Year, 1978-1988.

Year	No. of Projects	Total Production Cost (C$mil)	Spent in B.C. (C$mil)
1978	3	12.5	NA
1979	11	40.0	NA
1980	15	52.3	NA
1981	11	29.9	NA
1982	18	39.78	NA
1983	11	10	NA
1984	14	60	40.0
1985	28	150	76.0
1986	25	156	87.4
1987	28	282.9	152.2
1988	46	219.2	129.6

Source: Compiled from B.C. Film Commission, various sources.

[6]Different provinces in Canada work with different trigger ratios. For instance, the Alberta Motion Picture Development Corporation works with a trigger ratio of 2.5 times, whereas the B.C. Film Commission applies a ratio of 2.8. Alberta Film Commissioner Bill Marsden works with a multiplier of 4 for foreign film investment in his province, whereas in B.C. the highest multiplier used is 3.3. These officials differentiated between local investment and outside investment and gave the outside investment a higher multiplier. However, they could not explain why different ratios were applied by different provinces (Toth, personal communication, August 28, 1992).

Prior to 1984, the B.C. Film Commission did not identify whether all of a film's production cost was spent in the province.[7] It is also hard to delineate the sources of all of these investments from Table 10.2. These projects could have come from both Canadian-owned companies, as well as from companies in the United States and other parts of the world. The table, however, provides a useful account of the dollars invested in projects on film and how B.C. has succeeded in attracting a considerable portion of that investment.

IMPACT ON PRODUCTION

The trend since 1985 in overall production activity is clearly one of growth of all types of production in B.C. Feature films, movies of the week, TV pilots, and TV series shot in the province have grown (see Table 10.3). Although feature film production peaked at 16 films in 1988, it has remained at 14 to 15 in the following years, as Figure 10.2 indicates. The most dramatic growth, however, is in the category of TV series, from one series in 1984 to 15 in 1991.

Figure 10.3 breaks down the productions based on nationality, indicating the sources of investment in production in B.C. In the 13-year period, 1978-91, the overwhelming investment came from U.S.-based producers who chose to make their film and television shows in British Columbia. Only a quarter of the investment in this period was for Canadian productions, and about 3.3% went to productions of other nationalities (including co-productions). In 1991, the share of U.S. products being made in the province was even higher (76%; see figure 10.4).

Figure 10.5 compares film/TV production related expenditures in B.C. to those of total budgets. It provides a picture of overall economic impact from film and television production. Although there is considerable variance in film and TV production budget values from 1985 to 1990, there are corresponding but less drastic upturns and downturns in provincial expenditures over the same period, both of which steadily increase over time. The parallel patterns of film and TV and provincial expenditures suggest a long-term positive impact on the provincial economy.

---

[7]People in the industry usually refer to production cost as "negative cost." It is the total sum of money spent on preproduction (script preparation and planning, any costs related to the purchase of story material), all wages, production and postproduction (special effects, sound, and editing) expenses to complete the film. It does not include prints, publicity, distribution, and marketing expenses.

Table 10.3. Breakdown of All Production by Year, 1978-1991.

Year	Total	Features	TV Series	Movies Pilots	Drama
1978	5	3		1	1
1979	9	4	1	4	
1980	13	12		1	
1981	4	3		1	
1982	5	3		2	
1983	5	1		3	1
1984	9	3	1	5	
1985	24	10	3	11	
1986	25	8	4	13	
1987	30	6	9	15	
1988	33	15	8	7	3
1989	28	9	9	9	1
1990	39	12	11	13	3
1991	40	7	10	23	
Total	269	96	56	108	9

Source: In B.C. District Council, 1991 Report , Directors Guild of Canada, 1991.

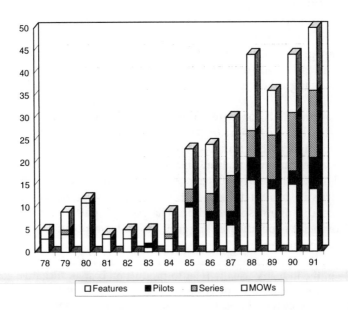

Figure 10.2. Number of productions in B.C. (1978-August 1991: By type)
Source: In B.C. & Yukon Council of Film Unions, 1991

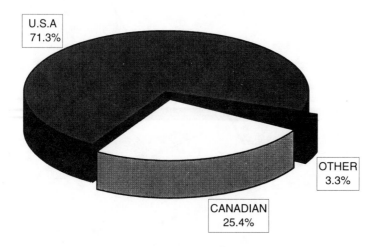

Figure 10.3. Productions in B.C. (1978-1991: Percent age by nationality)
Source: In B.C. & Yukon Council of Film Unions, 1991

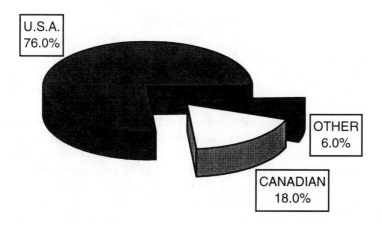

Figure 10.4. Productions in B.C. (1991: Percent age by nationality)
Source: In B.C. & Yukon Council of Film Unions, 1991

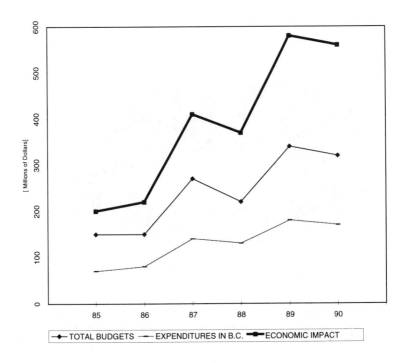

Figure 10.5. Production value in B.C. (1985-1990)
Source: In B.C. & Yukon Council of Film Unions, 1991

IMPACT ON JOBS

Every dollar spent in film production induces a multiplier effect on related economic activities. For instance, film production depends on a wide variety of supplies and services: sets, laboratories, special effects companies, dry cleaning services, transport, delivery, communication, food, hotel accommodations, construction workers, drivers, electricians, and many other technical and specialized labor. It is estimated that more than 1,500 retail and wholesale businesses in the Greater Vancouver area receive some portion of their annual revenue from film/TV production activity.

Bar graphs using 2.8 (Figure 10.6) and 3.3 (Figure 10.7) multipliers show total turnover measured in terms of dollars in related industries and services in the province. Figure 10.8 showing the cumulative economic impact aggregates the turnover from 1985 to 1990 and places it in the range of C$2 to $3 billion. All these data testify that B.C. has benefitted economically from the overall production activity that has occurred in the province from foreign and local sources.

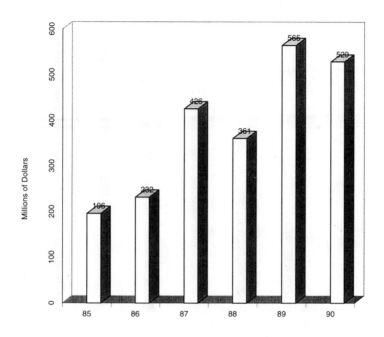

**Figure 10.6. Economic impact (1985-1990: Using 2.8 multiplier)
Source: In B.C. & Yukon Council of Film Unions, 1991**

Figure 10.9 visually demonstrates the positive impact of the production dollars spent in B.C. measured as growth in union workforce. These figures were drawn from the three major technical unions— Directors Guild of Canada, IA 891; Teamsters 155; and IA 669. IA 891 represents motion picture studio production technicians, covering a wide range of workers—accountants, construction, special effects, wardrobe, and so on. IA 669 includes camerapersons and other related technical personnel.[8] In the 11-year period tracked by this figure, union

[8]IATSE Local 659 in LA controlled the camera side of the local union, which was charted in 1919 in Vancouver. In 1981, the nationwide local 667 was formed in Canada to represent camera operators. In 1990, local 669 was granted a western charter to represent camera operators. Control over this union, however, remained in Toronto. Canadian locals of IATSE pay a percentage of their dues to the office of the Canadian V.P. of IATSE, who has a seat on the executive board of the International in the United States. George Chapman, president, IATSE local 891, told me that all Canadian locals are autonomous now, despite their links to the International (personal communication, September 3, 1992).

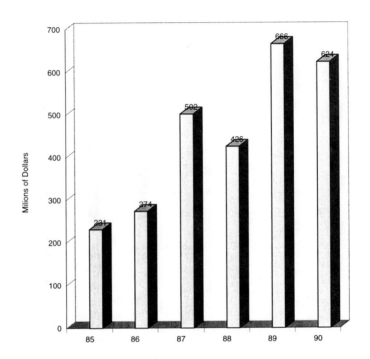

Figure 10.7. Economic impact (1985-1990: Using 3.3 multiplier)
Source: In B.C. & Yukon Council of Film Unions, 1991

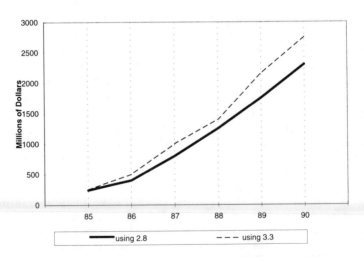

Figure 10.8. Cumulative economic impact (1985-1990)
Source: In B.C. & Yukon Council of Film Unions, 1991

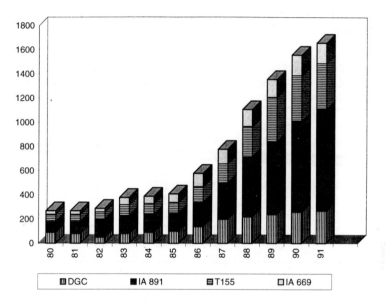

Figure 10.9. Growth in union workforce (1980-1991 excluding performers)
Source: In B.C. & Yukon Council of Film Unions, 1991

employment grew marginally in the early 1980s and dramatically since 1985. The graph excludes performers who may hold dual membership in the two unions representing them—the Association of Canadian Television and Radio Artists (ACTRA) and the Union of B.C. Performers (UBCP). UBCP claims to have 1,200 members and ACTRA 900.

The picture regarding employment is mixed when we compare the number of Canadian with foreign directors employed in these productions shot in B.C. (see Table 10.4). With the arrival of U.S. television producers in the province in 1985, the number of television series made rose by almost 100%. Canadians got 54% of the directorial jobs in local series as opposed to 46% in foreign productions. They appear to have gotten the lion's share of the series produced in 1987 and 1988, but their share shrunk quite dramatically by 1991 to 35% of foreign productions and 53% of Canadian series produced in the province. This essentially means that U.S. producers bring in members of the Directors Guild of America to direct the series rather than hire B.C. and other Canadian talent. One would have expected otherwise as, over the years, Canadians developed more expertise in directing U.S.-style TV series.

The Director's Guild of Canada (1991; B.C. District Council) expressed its concern over this issue eloquently:

Table 10.4. Breakdown of Series Percentages by Year and Director's Nationality, 1984-1991.

Year	Total Episodes	Cdn. Directors	Foreign Directors	Cdn. Production	Foreign Production	Cdn/Foreign Co-productions
1984	22	100%		100%		
1985	41	54%	46%	54%	46%	
1986	50	60%	40%	68%	32%	
1987	124	47%	53%	27%	73%	
1988	106	48%	52%	20%	71%	9%
1989	190	49%	51%	20%	57%	27%
1990	164	55%	45%	22%	51%	27%
1991	144	72%	28%	53%	35%	12%
Totals	841	56%	44%	42%	45%	13%

Source: B.C. District Council, 1991, Director's Guild of Canada, 1991.

> We in the DGC in B.C. are facing tremendous pressure in our collective bargaining to downsize the unit. It seems likely that in a year's time we could be facing the elimination in our contract of the provisions for minimum crew. Obviously, this may translate into fewer DGC members working on any given picture. . . . Add to this pressure of "offshore" Producers to import non-Canadian First Assistant Directors and the gravity of the situation is quite clear. (p. 3)

Both unions representing performers in B.C. appear to be overloaded. It is, however, difficult to say whether any of their members are getting opportunities to play key roles in the film and TV productions made in the province. The data obtained from the unions suggest that nearly 4,000 persons (actors, technicians, production personnel, ancillary service employees) were directly employed by the film/TV industry (Union of B.C. Performers, 1991). It is hard to tell, however, what portion of their annual income is drawn from the film/TV industry.

EXTERNAL AND INTERNAL REASONS FOR CAPITAL INFLOW

The primary factor driving the investment into B.C. appears to be the allure of getting more for less. It is a condition that is produced by both a state and labor pool that are flexible.

There are many external and internal reasons that have contributed to the movement of capital into B.C. When costs of production have escalated at home, the U.S. producers historically have taken their projects to places where they could stretch their dollars. Recipients of such investment are not only foreign countries but also those states in the United States where right-to-work (i.e., union-busting) laws were passed some years ago. The growth in U.S. investment in Canada coincides with the shrinking of feature film budgets in Hollywood, on the one hand, and the drastic audience loss to the U.S. networks on the other. Throughout this period of growth in production, the Canadian dollar remained weak relative to that of the U.S. British Columbia's diverse geographic beauty, comparable skill levels of technical labor, studio, hotel, and other infrastructure, its being in the same time zone as Los Angeles, and its being free of high crime and riots associated with parts of California, make the province desirable to U.S. producers. The ease with which Canadian cities can substitute in appearance for any city in the U.S. is also valuable to them.

There are some 200 state and city film commissions in North America that attempt to lure film and television production into their localities by offering to provide assistance of various types. B.C. appears to have an aggressive policy in recruiting production and in working with labor to establish favorable conditions for international producers.

The overall hospitable environment to foreign investors in film and television production in the country is also facilitated by the policies of the federal government. By 1988, the federal government had abandoned the Distribution Bill, aimed at remedying the problems in distribution of films made in Canada. A box office levy to raise funds for production was considered on several occasions, but never implemented out of fear of reprisals from the Hollywood cartel (MPEAA) and the U.S. government. The levies were demanded by the Canadian filmmakers as essential to strengthening the indigenous film industry. Consistent with its historic hands-off policy of U.S. control over distribution/exhibition in Canada, the government in 1988 created the Feature Film Distribution Fund managed by Telefilm Canada. The fund provided financial assistance to Canadian-owned distribution companies for acquiring rights to Canadian films and the better promotion of them (Telefilm Canada, 1990-1991).

Since its inception, the fund has disbursed C$49.2 million to a total of 22 different Canadian distributors. Of this amount, C$37.5 million supported acquisition of Canadian productions, $C11.8 million acquisition of non-U.S. productions, and $C2.4 million marketing and publicity by the distributors (*The Financial Times*, 1992, p. C3). The government spent another $9.9 million to support film festivals. Given the chronic undercapitalization of the Canadian-owned distribution companies, the government is subsidizing its operations by way of the Distribution Fund. The distributors, however, demanded that Canada pass legislation making it impossible for the U.S. majors to treat Canada as part of the U.S. domestic market, which is now the case. So far, that legislation has not been forthcoming. Instead, what has occurred is downsizing of national institutions (particularly the National Film Board of Canada and the Canadian Broadcasting Corporation) created to develop an indigenous film and television industry. In March 1992, the government reduced and froze the Telefilm Canada budget for the next four years at $144.2 million a year (*Calgary Herald*, 1992).

To encourage investors to participate in the film and television industries, the federal and state governments set up attractive tax writeoff incentives in the early 1980s. Until 1988, such certified productions received 100% tax writeoffs on federal income taxes over two years. As the Canadian economy plunged deeper into the recession, this writeoff scheme was revised starting in 1989 to allow only 30% writeoffs over two years.

More than 1,000 theatrical and TV films and series valued at almost C$2 billion have been certified as Canadian since 1985 by the Canadian Audio-Visual Certification Office in Ottawa. Certified productions are considered "national" productions and thereby become eligible for certain tax writeoff benefits. Earlier, I had argued that with this mechanism, the Canadian government was providing an indirect subsidy to the transnational corporations to make features and television shows in Canada (Pendakur, 1990). That is even more evident now, given the modifications in the certification criteria to allow international investments.

For productions to be eligible for certification, they had to receive 8 out of 10 points, allocated on the basis of nationality of the participant and the importance that person held in the production process (see Table 10.5). For example, a Canadian director, writer, producer, or principal actor immediately scored a point each. The rationale remains, but the current modifications may have watered it down. The 80% participation has been reduced to 75% to certify a production as "Canadian." Certain other concessions have been given as well that make it easier for U.S. residents of Canadian origin to qualify as being "nationals" for the purposes of the certification process. For example, actors such as Bill Murray, Christopher Plummer, and others originally from Canada who maintain permanent households in the United States would be qualified under the certification criteria.

At the provincial level, B.C., for instance, offers 30% rebates to Canadian investors in Canadian productions that have an international

Table 10.5. Certified Projects by Year and Cost, 1985-1990.

Year	No. of Projects	Cost (C$ mil)
1990 to Oct 31	176	375.0
1989	202	433.1
1988	164	487.5
1987	173	302.8
1986	199	235.0
1985	150	152.6
Total	1,002	1,986.0

Source: In Ruimy,1990, p. 58. © 1990 *Variety*. Reprinted with permission.

co-production partner. This rebate would apply toward that portion of the tax that is considered the domain of the province. The first production to get such an additional incentive was a feature, "Young Offenders." Michael Parker of Holiday Pictures set up a partnership with Shan Tam, a Hong Kong-based producer, to take advantage of the B.C. incentive. Another co-venture, "Arctic Blue," was done by Vancouver's New City Productions, Oxford Film Company of Seattle, and John Flock Productions of Los Angeles. The film, budgeted at C$7 million, had Canadian participation at C$2 million (Murray, 1993).

Relaxing the rules governing what constitutes "national" productions by the the Canadian Film and Video Certification Office (CAVCO) has encouraged other partnerships to form between international and local investors. "Adventure, Inc.," a two-hour movie pilot, was directed by Ted Kotcheff, who is a resident of California but for the purposes of CAVCO is Canadian. The film was produced by Toronto-based Primedia Productions and Spelling Television (Canada) for NBC and CanWest Global. It is the first project to come out of Spelling Productions, a leading Hollywood company, which established its branch plant, Spelling Television, in B.C. in 1992. As "Adventure, Inc." has the status of being a "national" production, it receives tax subsidies and rebates and becomes eligible to be counted as "Canadian" to meet the Canadian content regulations that were established by the federal government to encourage broadcasters to produce more local programs.

These policy moves—inaction on restructuring distribution and reduction in the budgets of national institutions—cleared any anxieties that Hollywood producers may have had about restrictive policies in Canada and paved the way for their increased investment. They are indicative of the general economic crisis in Canada throughout the 1980s and a lack of vision and leadership that has plagued government policy in the cultural arena for nearly half a century.

More decisive in encouraging U.S. companies to shoot in B.C. is the film unions' flexible policy regarding contracts with producers. At a time of shrinking economic opportunities at home, labor has become more malleable.

Two important contradictions in the labor unions shaped this policy of flexibility toward capital: regionalism versus nationalism, short-term versus long-term gains. The regionalism/nationalism issue can be understood best if we examine the jurisdictional disputes between two major unions representing actors—Association of Canadian Television and Radio Artists (ACTRA) and the Union of B.C. Performers (UBCP).

UBCP broke off from ACTRA when the local in B.C. raided a show called "Northwood" in 1990. ACTRA insisted that all contract negotiations had to go through the national office in Toronto in order to protect collective bargaining between actors and producers. The organizers of the breakaway union disagreed. The split arose from a tension within the industry over work opportunities in a highly competitive market, in which producers will set up production anywhere in the world to reduce their costs. The local argued that it had to make its own deal to protect its workers. Allan Krasnick (personal communication, August 31, 1992), director of collective bargaining at UBCP, was asked whether the formation of a breakaway union was motivated by economic or ideological demands. His response was:

> Well, it was both. One stemmed from the other. We were of the belief that unions had to be flexible in our contracts in order to attract foreign producers and keep them here in B.C. To make it attractive for them to pick B.C. instead of Portland, or Utah or some such place which is competing for production dollars, we had to remain competitive and flexible. This is the age of international production and companies will go wherever they can reduce costs of production. ACTRA was not going to be flexible like that.

In February 1990, five unions covering all crafts met to deal with work, enforceability, and pay. On all three issues, UBCP organized its membership, whereas ACTRA insisted all bargaining had to be done in Toronto. Krasnick said, "Our members here did not want Toronto to be telling them what was acceptable and what was not" (personal communication, August 31, 1992). The conflict appears to have been precipitated by Cannell Films, a U.S. company with a studio in Vancouver.

Krasnick gave the example of a TV movie to illustrate how that came about. If a movie is budgeted for $3.7 million, and the network fee comes to only $2.8 million, the production company will be short by almost a million just to break even. The only way it could recover that deficit plus other costs such as interest and overhead would be if and when the movie is sold to ancillary and international markets. In this situation, Cannell and other U.S. producers tried to reduce below-the-line labor costs, which is where the unions entered the picture.[9] Krasnick pointed out that in the Cannell movie contract, the unions in B.C. came in $120,000 better than Portland, OR in below-the-line costs. Concessions like these made them competitive with U.S. labor and independent of the national union leadership.

One gets the impression from talking with various union leaders in B.C. that they strategically worked with the B.C. Film Commission to attract foreign producers. George Chapman, president of International Alliance of Theatrical Stage Employees Local 891, which had about 1,000 members in 1992, told how he tried to lure international producers to shoot in B.C.:

In 1989, we went to Cannes for the first time. Cannes is a place to find distributors, but we were looking for producers. We got a list of all the attendees at Cannes and ticked off known producers from distributors. Checked the list with the Telefilm rep at Cannes and honed it down to a good list of producers who may be potentials to come to B.C. We went to their hotels and had the front desk people slip photocopies of invites into the producers' boxes. (personal communication, September 3, 1992)

Chapman stated that his union and Diane Neufeld of the B.C. Film Commission found a beach area for a reception, attended by 50 to 100 producers. The affair cost the union approximately $2,500, and, according to Chapman, "at least one producer came to shoot in B.C. due to this promotion. If he spent $4 million below-the-line for the shoot, then we service unions netted at least two million." He added that a similar reception hosted at the American Film Market in Los Angeles resulted in $18 to $20 million in business for B.C.

---

[9]The budgets are broken down into two major categories in feature film and television production. Above-the-line costs include wages and profit points (if any) paid to stars, directors, script writers, and producers. Below-the-line costs are all other expenses, including wages paid to technical labor, sets, raw stock, equipment rental, and postproduction. Generally speaking, in the films made by the majors, above-the-line costs account for a significant portion of the expenditures, sometimes up to 50% of the total negative cost.

The required outcome of such courting of foreign capital was flexible contracts. Krasnick of the UBCP stated, "We have had to constantly adjust our rates to the changing economic conditions of production on the one hand and competition from other states on the other. We have given certain concessions such as length of a work day, weekend over-pay, time and half and so on" (personal communication, August 31, 1992). Krasnick, however, insisted that UBCP rates were 28% higher than those of ACTRA.

At IATSE Local 891, Chapman echoed Krasnick regarding unions' flexibility to accommodate the marketplace. It essentially meant, according to him, "adjusting our rates to what our producers can pay and being realistic as to what the TV financing situation is like. We stole Jack's Place from LA. That is an ABC series. Signed a two-year contract with Spelling Productions. We have done the same for six years with Cannell Productions now. Adjusting to the market is what this is all about" (personal communication, September 3, 1992). Chapman said the most important agenda for his union is to create employment and not simply go after better wages and working conditions. He explained, "Our benefit program costs us $1,200 a member whether or not they work. So we have to be busy. Proactive. We don't sell memberships. We look at it as our obligation to provide work for our members. We are achieving it" (personal communication, September 3, 1992).

Such flexible contracts are possible because local unions have legal autonomy to negotiate contracts for themselves over the nationals,[10] and unions in B.C. appear to be evolving a more democratic governance structure. Krasnick of UBCP pointed out, "We are unique in North America. All contracts are ratified by the membership at large." The principal advantage of democratic governance is that if a contract signed with a company for a particular film ends up being undesirable for any reason, the membership at large can vote that company out when the next film comes along.

In such a highly competitive situation, B.C. labor can be pitted against that of other Canadian provinces, various U.S. states, or even foreign countries. In such a situation, the unions in B.C. have had to bring their rates down, according to Krasnick, who added:

> The companies will try to pit us against each other (other states and other countries). NBC (TV network) tried to do that with Night Tide. There is always that risk of competition and we have so far managed to

---

[10]Allan Krasnick suggested that this autonomy for locals has existed since the 1950s because of strikes in the construction industry. Jurisdictional conflicts between national and local unions were resolved favoring the local, thereby leaving governance at the provincial level.

stay ahead of competition. If we make a mistake, it is only with one production. Not for all films of a company. Besides, our membership can turn down the next contract that comes along. (personal communication, August 31, 1992)

IATSE Local 891 had approximately 1,000 members in 1992 and was growing at an average rate of 137 members a year. It was the largest craft union in the country, according to Chapman. In March 1992, IATSE Local 891 instituted a new governance structure of 23 departments, each with a representative on the executive committee. Seven other members of the executive are elected at large by the membership. The restructuring has meant that decision making in the union is more democratic, according to Chapman.

IMPLICATIONS

Because of the open arms policy and the financial incentives given by the federal and provincial governments, a significant expansion of the film and television production sector has occurred in Canada. The international producers who have flocked to B.C. basically churn out features, movies of the week, and TV series, all of which are principally directed at the U.S. market. The investors/producers are mostly non-Canadians, whereas some of the directors are well-known Canadians such Zale Dalen, Don Shebib, and others. These policies certainly appear to have created short-term opportunities for Canadian directorial talent.

At least three major differences exist between the "quota quickies" of the 1930s and the current film/TV productions financed by U.S. companies: first, in terms of their quality, current productions satisfy the industrial standards of the U.S. film and television markets; second, a sufficient number of trained personnel exist in Canada to provide the technical support for the foreign investors; and third, long-term investment by some of the U.S. producers is available. For example, Stephen J. Cannell, a leading California-based television production company, has built a large motion picture studio with seven sound stages in Vancouver. Its subsidiary, Cannell Films of Canada, Ltd., has produced "Wiseguy," "21 Jump Street," "JJ Starbuck," and "Commish," all of which have aired on major U.S. television networks during primetime. Another prominent producer, Spelling Productions of Los Angeles, has also established a production facility in Vancouver. Such branch plant activity in B.C., of considerable size and importance, has resulted primarily because of rivalry among unions and cost-reducing incentives.

Canada still makes the small-budget, indigenous films funded by a consortium of federal/local state agencies, private investors, and so on. These films are not necessarily driven by the market but by the vision of independent artists. They are usually not capital-intensive, unlike those that are destined for the U.S. market, but are distinguished by someone's passion to tell a story from a personal point of view or to innovate and push the boundaries of the art form. Such filmmakers have survived across Canada despite the flood of investment into the commercial sector, emerging in the shadow of the U.S. film and television industry.

Labor negotiations with international producers and the rates at which they work for those productions do not seem to affect small-budget, locally made Canadian films. Two directors who have recently done such features, Srinivas Krishna ("Masala," 1989) and Atom Egoyan ("Calendar," 1993), said they work under a very different set of production conditions. They hire established cinematographers, editors, and others with whom the directors maintain long-term relations. Egoyan said that his crew members know the limitations of his budget and understand that he cannot pay them the kind of wages they can get from features made by U.S. producers.[11] Films such as "Jesus de Montreal," "Adjuster," "La Demoiselle Sauvage," and "Highway" will get made as long as the support structures of the government of Canada and various provinces remain intact.

In the short run, the strategies pursued by the unions in B.C. appear advantageous to their immediate goals of creating employment in a crisis-ridden economy. Given the changes in the global economy in the last decade and the liberalization regimes at work in many economies around the world, organized labor faces a critical dilemma of choosing between long-term needs and short-term opportunities. B.C. unions have chosen the former, and their leadership is not certain whether this is the right path.

No one, however, knows how long the production boom will last as it hinges on many interrelated factors, as this chapter has attempted to show. Any attempt at prediction is risky. If one considers market expansion and demand factors in the United States as key variables to Canada's relative prosperity as an offshore, branch plant for Hollywood, it is safe to say that Canada will be attractive as long as it remains competitive in labor costs. Other U.S. phenomena may create more opportunities for Canadian labor. Major technological changes (digital compression) are reshaping the cable delivery system in the

---

[11]I spoke to Atom Egoyan and Srinivas Krishna at the "Alternative Visions: Festival of New Canadian Cinema," held at the Program on Communication and Development Studies, Northwestern University, Evanston, IL, April 20-23, 1993.

United States. Additionally, the political-economic conditions are spurring significant changes in the structure of telecommunications and entertainment industries. Megamergers and joint ventures (e.g., Fox-TCI, Warner-US West, MCI-NewsCorp) between once rival corporations are collapsing the older divisions between the movie industry and cable/telephone industries. Such collaborations of big capital in the United States are geared to enhancing the number of channels and offering more pay-cable, pay-per-view choices. If those planned operations materialize and a sufficient number of upscale households sign up, the result could be higher demand for film/TV products.

Neither Canadian nor international unions appear to be ready to negotiate with the emerging entertainment industry. They are caught up in the daily demands of survival in which narrow economic interests set the agenda. In the final analysis, workers are pitted against each other: local against local, national versus the provincial, and so on. International labor solidarity, in which autonomy of the locals is preserved and democratic governance maintained, needs to be reconsidered in these changed conditions. Control over one's creativity, a tough issue that the founders of the Screen Writers Guild grappled with in the 1930s, must be reconsidered in order to put the item of division of labor back on the agenda in Hollywood.

Jeff Kibre, a nonconformist Communist labor organizer of the 1930s, wrote, "within this world of fantasy are thousands of workers of all kinds who sweat, and who fight the same bitter struggles as other thousands of workers to secure a decent livelihood" (quoted in Ceplair, 1989, p. 64). At this moment in history, we need hundreds of organizers like Jeff Kibre who worked for raising the collective consciousness of workers in Hollywood rather than the craft vocation and economism pursued by labor organizations in the industry.

REFERENCES

Adilman, S. (1990, November 19). French flicks reap bucks. *Variety*, p. 56.

B.C. & Yukon Council of Film Unions. (1991, August 23). *A graphical representation of the growth of the British Columbia Film & Television Industry*. British Columbia: Author.

Calgary Herald. (1992, March 26). p. C5.

Casselton, V. (1993, February 26). Lost business convinces unions to hitch up and hype B.C. *The Vancouver Sun*, p. D1.

Ceplair, L., & Englund, S. (1979). *The inquisition in Hollywood: Politics in the film community, 1930-1960*. Los Angeles: University of California Press.

Ceplair, L. (1989). A Communist labor organizer in Hollywood: Jeff Kibre challenges the IATSE, 1937-1939. *The Velvet Light Trap, 23,* 64-74.

Director's Guild of Canada. (1991, February 14). *B.C. District Council, 1991 Report.* British Columbia: Author.

Evans, M. (1988, May 7-13). U.S.-based productions boost local film industry. *TV Week, 12*(19), 4.

Financial Times. (1992, June 17). p. C3.

Hartsough, D. (1989). Crime pays: The studios' labor deals in the 1930s. *The Velvet Light Trap, 23,* 50-63.

Hollywood Reporter. (1992, April 17). pp. 1, 121.

Hot Property (n.d.). *Quick facts about the B.C. film industry.* British Columbia: B.C. Film Commission, Vancouver.

Hot Property. (1992, March). *Production and budget guide.* British Columbia: B.C. Film Commission, Vancouver.

Huettig, M.D. (1944). *Economic control of the motion picture industry: A study in industrial organization.* Philadelphia: The University of Pennsylvania Press.

Movie makers may get break on fees. (1993, January 19). *Toronto Star,* p. 16.

Murray, K. (1993, May 10). B.C. on way to new peak. *Variety,* p. 6.

Pendakur, M. (1990). *Canadian dreams and American control: The political economy of the Canadian film industry.* Detroit: Wayne State University Press.

Ruimy, J. (1990, November 19). Triple threat to Canadian pics. *Variety,* p. 6.

Telefilm Canada. (1990-1991). *Feature film distribution fund. Policies, 1990-1991.*

Union of B.C. Performers (UBCP). (1991, September 23). *$600 Million in B.C.'s film industry.* British Columbia: Author.

11

The Animation Industry and Its Offshore Factories

John A. Lent

In Colombo, Dhaka, Bombay, in city after city throughout Asia, cartoonists are searching for ways to get into animation. Not so much domestic animation, which would help fill program slots on the proliferating television channels in their countries, but rather foreign series. They want to follow the examples of South Korea, Australia, Taiwan, and the Philippines, where local branches of United States and European animation studios have been established to produce "The Simpsons," "Ninja Turtles," and a myriad of other titles. They contend that they can underbid these countries in the resource in which foreign companies are most interested—a cheap and stable workforce. They are willing to work directly with the Western companies or to act as subcontractors for their Asian branches.

A few reasons account for this phenomenon prevalent in Asia, Eastern Europe, and Australia, chief of which is the renaissance of animation in recent years. Thousands of new television and cable channels have emerged as a result of the rush to privatization and satellite hookups and in response to the public's interest in, and infatuation with, animation. The public response is evident from the new all-cartoon networks, primetime animation shows, and increased time slots on weekday afternoons and Saturday mornings allotted to cartoons. Commenting on this interest, Kurcfeld (n.d.) said:

> In an age where politics comes in speech balloons, human psychology in slim volumes at the checkout counter and we all have three-minute attention spans, where advertising and showbiz have infantilized the populace, and everything is slick and simple, cartoons may become the most realistic illusion. (p. 32)

An outcome of the demand for more animation has been a shortage of animators in the United States. Deneroff (1994) reported that of the 1,600 active members of the Motion Picture Screen Cartoonists, only 100 were unemployed, and that it was probable that even they were working for nonunion shops. The same talent shortage was reported across North America and Europe. In the United States, according to Deneroff, major studios hoping to recruit and retain class-A animators offered up to $2,200 a week, twice the union minimum. He said that their inability to find talent and the augmented budgetary problems were the factors that forced U.S. studios to send more work abroad.

Such an analysis is too simplistic and not completely accurate. U.S. studios, much to the annoyance of the animators, have been farming out work for more than a quarter of a century, and the relationship between producers and animators has fallen far short of harmonious. In 1979, animators in the U.S. walked out over "runaway production," the sending of animation work abroad and depriving U.S. cartoonists of work. The resolution of the strike between the producers and the Motion Picture Screen Cartoonists IATSE Local 839 was a clause in the contract that prohibited the studios from exporting any work from Los Angeles County until qualified union rank and file were hired.

The union tried enforcing the clause to the extent that a $200,000 fine was levied on Hanna-Barbera for using South Korean and Taiwanese labor, but the studio was not deterred, continuing to send two thirds of its work to those countries in 1982. That same year, the producers sought to delete the clause restricting the use of foreign labor. The concerns were real on both sides: The union traditionally experienced high unemployment (40% in 1982) and seasonal work (often only 22 weeks a year); the producers, high production costs. A major reason they sought foreign labor was to avoid paying fringe benefits, which, they claimed, in the United States added a 32% cost factor beyond wages (Goldrich, 1982).

LABOR EXPLOITATION

U.S. animation was founded on labor exploitation from the beginning, as talented animators worked extremely long hours at grueling, tedious jobs for low wages and with virtually no credit. In the formative years of the animation industry (1920s and 1930s), labor was cheap and plentiful and unions unthinkable. As Klein (1993) reported, until the late 1930s, the salaries were about the lowest in the film industry, "essentially comparable to clerical and construction wages in the country at large" (p. 93). At Disney, salaries for inkers and in-betweeners (those who draw parts of cartoons between sketches and finished product) ranged from $17 to $26 a week in the 1930s and early 1940s. The hours of work and footage quotas (oftentimes 4,000 drawings per worker per month) required of animators were staggering, reflected in an inside joke found on a watercolor done by a Disney employee. Shown in a corner is a slumped-over animator, while Walt Disney is on his knees praying, "Please God, send me an animator who can work twenty-four hours a day" (p. 97). Disney animators (and presumably those at other studios) took work home to polish and were expected to donate a half day on Saturday without pay and a few evenings each week for art classes.

Added to all this was the fact that talented artists/animators felt "utterly marginalized within the studio system (treated and paid like craft workers who did not make movies, only produced a related service)" (Klein, 1993, p. 183). At Warner Brothers and MGM, animators were neglected, isolated from the rest of the studio business and made to endure "effronteries" from their bosses, Leon Schlesinger or Fred Quimby; at Disney, they were also treated poorly, even spied on, as Walt Disney helped set up the McCarthy era Hollywood blacklist. Many creative geniuses (Otto Messmer, Tex Avery, Carl Barks) remained anonymous, as their producers (Schlesinger, Quimby, Disney, or Sullivan) grabbed the spotlight. Some of these talents were not known to the public until the 1980s and 1990s, when revisionist histories of the industry were written. One of Avery's gagmen of the 1940s described how indifferently Avery had been treated:

> Five hundred dollars' worth of jokes in a minute. Unbelievable. The most unbelievable thing was that they didn't appreciate it, that they didn't snare him and elevate him to the papacy of humor on the front lot. Tex could have been a Frank Capra or whatever in this business, could have gone completely to the top. The cartoon business is full of brilliant people like that who never get heard of. (quoted in Klein, 1993, p. 185)

Another concern of the animators had to do with how their work was treated in the more rationalized, factory-style production mode that the Bray Studios introduced in the 1920s. Klein (1993), calling it Fordist or Taylorist, described this way of doing animation:

> Good business required that the work of four to six animators be sewn into a single cartoon. . . . It had to be very orderly, an order that produced its own problems: the tendency to regiment the product. . . . Increasingly, as the pyramid of workers grew, the head of studio became the chief story man, with many bottle washers below. . . . The many stages, with so many more hands at each stage, required enormous agreement as to how the style had to survive to the end. At least five different people would redraw and paint the very same images. (p. 89)

As the studios expanded their employee base, continuing characters went through many more hands ("retraced to death"; Klein, 1993, p. 104), with the result that they sometimes became too routinized—less spontaneous, surprising, and funny.

By the 1930s, labor unrest was felt at different studios, especially at the lower levels, from inkers to assistant animators. Fleischer Studio first experienced a near-walkout in 1935, before its long, brutal, and bitter strike of 1937. The major strike issue at Fleischer developed with the creation of a labor glut when the neighboring Van Beuren Studio went under, keeping wages to a minimum and in some cases cutting them to below the cost of living in New York City, then home to those studios. The low pay scale at Disney made animators desperate for a union contract in 1940. This led to a strike a year later and provoked Walt Disney's vindictive reaction in the firing of union organizers on his art staff and labeling them Communists. The Schlesinger and MGM studios were barely able to avert strikes; Warner was not so persuasive, as employees walked out both in 1945 and 1946.

Job security became a major issue after 1953, when the entire industry faced a crisis. The number of cartoon play dates dropped dramatically, production and exhibition costs skyrocketed (between 1941 and 1956, production costs rose 225% and rentals only 15%), and, by 1963, companies such as Warner Brothers, UPA, and MGM were forced to bow out of the production of animation shorts. As Klein (1993) pointed out, "Consumer marketing changed public habits considerably, particularly after the collapse of movie chains and the arrival of the topsy-turvy world of television" (p. 240). As a result, the big studios cut spending on marginalized areas such as cartoons when they found they could re-release old cartoons at 90% of the earnings of making a new one.

Those studios remaining shifted their work to television, where many of the animators had gone, often to do commercials. Bill Hanna and Joe Barbera were among the latter. When MGM closed their cartoon studio and laid off all their animators, Hanna and Barbera found a spot for their character, "Ruff and Ready," in the 1957 television schedule. From there, they produced at least 138 TV cartoon series over the next 30 years. They also revolutionized production by converting from full to limited animation and, eventually, sending much of the work overseas. The export of labor followed the trend of many U.S. industries in the 1960s of shifting to offshore production.

GOING GLOBAL

Hollywood's first offshore animation was done by Japan in the early 1960s, a natural choice because Japan was one of the few Asian countries that had an animation industry. China had well-established animation, dating to work finished by the Wan brothers in 1926, but because of diplomatic restrictions, it was off limits to U.S. business. Although India produced an animated film as early as 1915 and North Korea an animation studio in 1948, neither country had adequately developed the industry.

In the 1960s and early 1970s, other countries joined Japan in the production of foreign animation, notably Taiwan, South Korea, and Australia. Ying Jen Advertising Company in Taipei, in conjunction with a Japanese animation firm, trained animators who for one year (1968) processed Japanese TV cartoon series. U.S. animation was first done in South Korea in 1969; four years later, the Japanese company, Golden Bell, had animation done in Seoul.

South Korean entrepreneurs were quick to see the potential of offshore animation, as they picked up U.S. cartoon skills and launched studios in the 1970s. The oldest surviving animation company, Dai Won, was started in January 1974 by animator Jung Wook, mainly to handle Japanese contracts. In 1985, Dai Won did the American "G.I. Joe" series under Japanese subcontract (Jung Wook, personal interview, August 14, 1995). In the late 1970s, Tayk Kim, through his Dong Seo Studio, processed first theater and then television animation for the United States. He claimed that in the 1970s and 1980s, most South Korean animation companies worked on a subcontracting basis for the Japanese, and that his studio was one of the rare ones dealing directly with the United States (Tayk Kim, personal interview, August 16, 1995). A third pioneer in foreign animation production is Nelson Shin, whose AKOM is one of the largest South Korean studios today, holding many contracts, including one for "The Simpsons." In 1980-1981, Shin met at the recently established Marvel Productions with executives who wanted to know if he could produce for them a 75-minute feature within two months. "I said let me go back to Korea and figure it out," Shin recalled. By hiring "almost all the Korean animators," Shin managed to complete it within 10 weeks, after which, a lot more U.S. business flowed his way (Nelson Shin, personal interview, August 16, 1995).

Australian animation started in the 1960s with productions of "The Beatles" and "Beetle Bailey" for U.S. television, using Australian talent. As one writer commented, "this relationship between Australian talent and the U.S. networks has been the cornerstone of our industry" (Balnaves, 1996, p. 123).

Actually, Australia laid claim to having the first major offshore studio for any U.S. animation company, when Hanna-Barbera opened its unit there in 1974. During the next 15 years, the Hanna-Barbera Australia studio produced more than 500 half-hour segments of animation. In 1989, the studio was sold to Disney, which has made its series, specials, and made-for-video movies there, at the same time providing employment for a major part of the Australian animators' pool. Most Australian studios trace their heritage to Hanna-Barbera, one of the main ones being the Southern Star Group, with its divisions, Mr. Big Studios and Southern Star Pacific. Mr. Big Studios consistently supplies Hanna-Barbera and Nickelodeon.

In 1978, Hanna-Barbera helped set up another major offshore studio, this time in Taipei. Its founder was James (Chung Yuan) Wang, a Taiwanese who had studied film in the United States. While working in the Los Angeles office of Hanna-Barbera, Wang submitted a proposal for what became known as Cuckoo's Nest (also known as Wang Film Production Co., Ltd.), and it was accepted. One of the world's largest production houses, Cuckoo's Nest has grown from a basement studio to two multistoried complexes encompassing 150,000 square feet, from a handful of staff to more than 900 employees, from an initial investment of US$330,000 to annual revenues in excess of $27 million, and from an output of 17 cartoon episodes to a peak of more than 200 yearly. Cuckoo's Nest collaborates with Disney, Warner Brothers, Hanna-Barbera, MGM, Nelvana, CBS, and France Animation, as well as other U.S., French, Canadian, Japanese, and German companies. Nearly half of all cartoon shows on U.S. television credit Cuckoo's Nest (James Wang, personal interview, July 10, 1992). When labor costs escalated in Taiwan, Wang set up branches in other countries, with major involvements in Thailand and China. Thai Wang has about 600 staff members in seven or eight studios, whereas the China facility in Guangdong employs 200 to 300.

THE HEADQUARTER COUNTRIES

The economics of the industry made it feasible for Asia to feed the cartoon world. Cheap labor abounds in the region, and U.S. animation executives readily acknowledge the cost factor in sending work abroad. According to Phil Roman, owner of Film Roman, the largest independent animation studio in the U.S., "If we had to do animation here, it would cost a million dollars instead of $100,000 to $150,000 to produce a half hour, and nobody could afford to do it except for Disney" (Edelstein, 1995, p. 38).

The price is right also because overseas producers can provide a large pool of cost-efficient, labor-intensive jobs in drawing and coloring. Although such work can be done by computers, the finished product suffers in comparison to hand-painted cels. John Kafka, vice president of Bobtown, a Los Angeles production house serving as a go-between for U.S. studios and foreign companies, complained that hardware used to depict motion comes up short in "delivering the lampoon-like elasticity of hand-drawn animation"; basically, the result is not as funny.

The highly motivated, productive, and stable nature of Asian labor is equally a key factor. James Wang of Cuckoo's Nest said, "People in Asia work very hard. Getting people to work well here is easier than elsewhere. Labor is still high quality. This area of the world is still influenced by the Confucian philosophy on work habits" (James Wang, personal interview, July 10, 1992). At the Wang branch studio in southern China, animation employees work seven days a week and are paid overtime; their only time off is a short vacation at Lunar New Year. Because of the shortage of job opportunities and the severe need for employment, and, in some cases, antiunion or antistrike legislation, Asian animation workers are not likely to strike or cause upheavals and stoppages. Nevertheless, U.S. companies safeguard their overseas operations by diversifying production places, thus ensuring that work continues if trouble does occur. For example, Film Roman alternates the production load among six different overseas studios, all in South Korea. After a studio is chosen, it receives the layouts for an episode and is paid roughly one tenth of the total budget (about $120,000 to $160,000) for a nine-week job.

The usual procedure is for pre-production to be done in the United States. This consists of preparing the script, storyboard, and exposure sheets that determine timing and spacing of a show. In the next stage, the package is sent overseas for production, where the drawing of cels, coloring by hand, inking, painting, and camera work are completed. The work is sent back to the United States for postproduction (film editing, color timing, and sound).

The United States is dominant in offshore animation, although Japan and some European countries also are involved. Disney does much of its work overseas, either through its own animation companies in Japan and other countries or as the sole client of a domestically owned studio such as in South Korea and Canada. When Disney expanded its production of theatrical and television cartoons in 1984, the studio turned to Japan for labor. By April 1989, all Japanese companies producing for Disney were streamlined into the newly established Walt Disney Animation Japan. At this Tokyo branch, Japanese animators drew all pictures (25,000 for a half-hour program, plus 400 more for background), inked, colored (using more than 300 colors, all designated in the United States), and shot the drawings on 35-mm film, which was then personally carried to the United States. Because about 1,000 people are needed to complete a program and given Japan's labor shortage, Disney Japan by the early 1990s was subcontracting more than half of the labor to South Korea or China. As Masaki Iizuka, Disney president in Japan, said:

> Because of its labor shortage, America has turned to . . . Japan and Australia for production, but Japan in turn has a labor shortage. Eventually, we've ended up having to ask every studio in Asia. This business simply does not have borders as it is being set up as an international business. (quoted in "Made in Japan," 1990, p. B-16)

As noted earlier, Japanese companies producing their own shows also have resorted to using foreign labor. About 20 firms directly produce animation films, although numerous smaller studios also are involved. To cut costs, and because of a labor shortage in Japan, a large proportion of the production work is done in the Philippines, South Korea, Taiwan, Singapore, and Thailand. For example, one of the large companies, Toei Animation, has had a joint-venture subsidiary in the Philippines since 1986 and a partnership in South Korea even longer. The Japanese companies tend to follow the U.S. studio approach, in which planning, basic character, and background drawings are done at home and inking, painting, and main quality control abroad. Increasingly, the Japanese have been farming out background drawing as well (Feazel, Galbraith, Demuth, & Gwynne, 1991, p. 15).

Much of the work in European animation—the majority of inking and drawing, in fact—is sent to places such as South Korea, North Korea, or Vietnam. European animation companies operate differently from those in the United States, with many co-productions in evidence. Nonetheless, animation is too expensive for most financiers, and animators must look outside Europe for investment and for labor (Walker, 1994).

The European goal has been to bring the work back home. Some companies such as France Animation of Paris have complained of increased delays, high labor prices, and inconsistent quality, as Asian subcontractors are overwhelmed with work from dozens of international production houses (Williams, 1991). By 1995, there was a noticeable decline in the work subcontracted outside Europe, partly because of a big increase in the use of computerized animation systems such as Animo or Tic Tac Toon and because of the integration of Europe as a single market, resulting in more cross-border co-productions. Already operative are more than a dozen European studio groupings, each combining studios from at least three different European countries. An example is the Anixa grouping, made up of Telemagination of Britain, Praximos of France, and NFP Animation Film of Germany. The inexpensive labor pool for these groupings comes out of Spain and Eastern and Central Europe. Russian studios, for example, are finding more opportunities to co-produce with the West, offering high-quality animation skills in exchange for an equity position (Swain, 1995). All this is not to say that European animation is no longer visible in Asian cartoon factories. In South Korea, Taiwan, and China, many studios have European contracts and will continue to have them, as Asian subcontractors "still have a cheap manpower [situation] that can work with the latest Western hardware and software" (Rodriguez, 1996, p. 130).

THE PROCESSING COUNTRIES

Asian animation companies continue to bid fiercely for parts of the global business, insisting that they provide employment and skills for young people, bring in needed foreign capital, and add to the creation or enhancement of domestic animation.

Most animation workstations in Asia are, in fact, staffed by young people, many of whom are women. At Seoul Movie Co., Ltd., for example, 80% of the 300 employees are women, ostensibly hired for their "delicate touch"; their age range is from early 20s to 40s (Jun Chang Rok, personal interview, August 14, 1995). About 35% of the 500 employees at Seoul's HanhoHeung-Up Co. are in their 20s. Similar characteristics are found in the other offshore animation factories. At South Korea's Dai Won, 90% of all assistant animators are in their 20s, having joined the company directly from high school; at Shanghai Yilimei Animation Co., most workers are 30 years old or younger, and women make up about 35% to 40% of the key animator slots. The majority of the Wang staff in Taipei are under 30 years of age, one source giving the average age as 22 to 23.

Animation officials and, in some instances, the workers themselves declare that their wages are competitive with other sectors of the economy. The head of HanhoHeung-Up said that at one time wages in animation far exceeded those in other fields, but that was no longer the case. In the busy season, a painter or inker at his studio reportedly can make US$1,000 a month and receive medical benefits, four days annual vacation, one day per month guaranteed legal vacation, and a retirement pension (Kim Seok-ki, personal interview, August 13, 1995). At Seoul Movie Co., Ltd., salaries are said to range from about US$1,000 per month for the lowest colorist and cel cleaner and from about $4,000 to more than $13,000 monthly for upper-level animators. Such salaries compare very favorably with university graduates' starting salaries of $1,000 a month. Additionally, medical insurance, accident security, and other fringe benefits are provided (Jun Chang Rok, personal interview, August 14, 1995). Inkers and painters at AKOM, South Korea's largest independent animation house, are paid $1 to $1.50 per cel; the rate is a bit higher if the work is by hand, rather than by computer. During AKOM's busy season, 1,200 animators are hired, 1,100 of them temporary in anticipation of eventual downtime (Nelson Shin, personal interview, August 16, 1995).

The Philippine Animation Studio Inc. boasts high wages tied to production quotas. An efficient PASI animator is expected to do 50 feet (35 seconds of television) of drawings weekly, which reportedly can earn him or her the huge sum (by Philippine standards) of $2,000 to $2,700 a month. However, to earn high salaries, technicians who fill in and color animators' drawings virtually live at the studio, sleeping on cots next to their desks or in bunk beds provided by the company (Karp, 1995).

At another large Philippine studio, Fil-Cartoons, the base rate for an animator is said to be $200 per week and with incentives, $500 to $1,000. However, the wage for inkers and painters, according to this source, is less than $100 weekly. Fil-Cartoons, in which Hanna-Barbera has a large financial interest, has expended a great deal of energy to improve working conditions and morale in recent years. Highly publicized celebrations (such as when the company produced its one millionth foot of animation in March 1995) and ceremonies (such as "An Affair To Remember," honoring employees with five or more years of service) have been added to less splashy perks such as a better cafeteria menu, a company newsletter, an emergency loan fund, free lodging for those working mandatory overtime after midnight, and classes for staff members ("Cartoon news" 1995, p. 8).

Wang Film Production Co. in Taiwan also takes pride in how it treats employees. In fact, one of the three points of the company code states that "our employees, in return for their loyalty and productivity, deserve and should expect the loyalty and respect of Cuckoo's Nest Studio, and a stable, secure working environment in which they will be treated fairly and individually." James Wang believes one way of satisfying, and therefore retaining, employees is to keep their skills highly honed through many training classes. "You cannot force people to work, but if you train them, they feel they are not being used. Here, people are number one, and money is number two. This way, you can create ideas" (James Wang, personal interview, July 16, 1992). Wang also arranges two company trips yearly, provides dormitory space, and pays a competitive wage. Major animators reportedly draw a salary of $1,900 a month, 60% of their U.S. counterparts.

Even in situations in which wages are low and working conditions poor, some people in Asian animation rationalize away the labor exploitation factor. Ram Mohan, head of Bombay studio bearing his name, said: "I don't see it as such [exploitation]. From an Indian perspective, if I do animation for Hollywood, it is an opportunity for young people to find a career. There are very few chances for artists; this would open up a large area of employment" (Ram Mohan, personal interview, July 10, 1993). The vice general manager of the Shanghai Yilimei Animation Co. also said he could not see how it was labor exploitation. According to him,

Comparing with companies involved in foreign investment of any type, we are above median and lower than the highest in salaries. I am not familiar with others' salaries, but our animators' pay is close to that of professional artists generally. The salary is not fixed; it depends on how one produces. Piece rate. It depends on how many projects we have with foreigners as to whether there is enough work to keep employees well paid. (Jin Guo Ping, personal interview, August 13, 1993)

A young animator, who has worked for both the state animation house and an offshore facility, felt he had been exploited at both. He told how the Shanghai Yilimei Animation Co. had given him only 800 yuan (at the time about $160) above his salary for an award-winning show he had labored on for a year; yet, he used to receive more than six times that amount monthly while working for the offshore plant, Pacific Rim. He added:

Today, they are paid 8,000 to 10,000 yuan [about $1,600 to $2,000] a month working for Pacific Rim, but it is brainless work. The workers are exploited. When they asked the boss for more pay, he said okay, but you have to work harder. They stopped requesting pay increases. Our cartoonists don't think of this as exploitation. Instead, the youngsters judge by the money they make from foreign compared to local work. (Zou Qin, personal interview, August 13, 1993)

In 1995, Pacific Rim closed, citing financial problems brought on by workers' constant demands for higher pay (Thornton, 1995).

In places where freelance help is readily available, animators face a particularly precarious existence. The general manager of Run Win Arts Zhuhai, a Wang subsidiary in southern China, said he keeps workers in line by warning them that they can be quickly replaced. He called this the Chinese style of managing, adding, "when my animators ask for higher wages, I just remind them that I already use freelancers for about 10% of our work. I won't pay a higher wage, but I might freelance more of our cartoons" (Thornton, 1995, p. 86).

Governments of the region occasionally look kindly on the foreign animation work, mainly because it brings in foreign exchange. In 1994-1995, the South Korean authorities, impressed with the large amount of foreign capital offshore animation companies attracted, granted the industry a favored "manufacturing" status and co-sponsored a huge animation festival. Ram Mohan said the Indian government also is fully supportive of ventures such as foreign animation processing. Normally, he said, the export of celluloid materials from India is expensive, but if one is able to get the status of an export-oriented unit— exclusively for work abroad—the government grants concessions (Ram Mohan, personal interview, July 10, 1993).

A claim often heard is that offshore animation leads to the creating and nurturing of a local industry as an infrastructure is built up, equipment is put into place, and skills and know-how are transferred. In some cases, animators say they work for the foreign companies to gain an edge in setting up domestic animation operations. The manager of Kantana of Thailand, which does work for Japan's Toei, said, "We have plans to produce our own animated comics. That was why we agreed to work for Toei. [By working with Toei,] our cartoonists can learn more about animation techniques. . . . Toei even provides a training course for the cartoonists" (Sirimas, 1993, p. B-1).

With the explosion of television and cable channels in Asia, stemming largely from privatization and the recognition of cartoons as an adult and not solely children's medium, domestic animation has taken on a status it never before had. Perhaps the country moving most rapidly to establish a local animation industry is South Korea; this is due in large measure to government initiatives. In 1994, government officials, including the president, and technocrats recognized that animation is a "value-added" product for South Korea; the government has since provided tax and other support in the form of low-interest loans for the industry. These actions, in turn, have encouraged more domestic production and the launching of about a dozen junior college and university animation programs. Even before the government moves, Dai

Won devoted 40% of its production to domestic shows, and Seoul Movie Co., Ltd. also had a large domestic production schedule. Other companies that were almost solely processors for U.S. and other animators are raising the number of local shows in production. HanhoHeung-Up, which already obtains 10% to 15% of its revenues from domestic production, plans to increase that total, and AKOM, totally immersed in foreign work, expects to do domestic animation once it becomes profitable. By 1996, a few feature-length and numerous shorts depicting Korean folklore, humor, and culture were on the market.

Concentration on domestic animation has not been as keen in other Asian countries. In most cases, animators were proud when they could point to one or two features that had local angles. Wang Film Production, during its first 15 or more years, had virtually nothing to show in the way of domestic programs; however, in July 1995 alone, the company launched three 33-minute episodes of "The Little Monk" and a one-hour feature, "The God, the Taoist and the Goblin." In Thailand, Kantana Animation, a workhouse for Japan's Toei, produced the first-ever Thai television series in 1994 and started a feature, "Mome the Dog," while Fil-Cartoons in Manila created "Swamp and Tad" in 1995. The latter already has been bought by both the U.S. and Asian cartoon networks.

Of course, Shanghai Animation Film Studio produced only Chinese animation for decades until its conversion into Shanghai Yilimei Animation Co., a joint venture with Hong Kong's Yick Lee Development Co., Ltd. (Segers & Michie 1988). However, according to the vice general manager, as a merged company, "we have to compete with other locals working for foreign companies and still pay attention to our original purpose of educating our children" (Jin Guo Ping, personal interview, August 13, 1993). Seventy percent of Shanghai Yilimei's funding still comes from government, allocated on the basis that the company completes a 300 to 400 annual minutes quota set by the authorities. Any money needed for development and creativity comes from the 500 to 700 annual minutes of animation done for foreign firms. Jin Guo Ping (personal interview, August 13, 1993) explained that foreign and domestic animation will parallel each other in China: The domestic will be for the artist, the foreign for commercialism.

South Korean animation executive Jun Chang Rok (personal interview, August 14, 1995) also saw close links between domestic and foreign animation. He said his Seoul Movie Co., Ltd. must focus on commercial animation for foreign clients to obtain the monies needed to reinvest in the artistic pursuit of domestic animation.

CONCLUSION

Offshore animation is too recent a phenomenon for one to easily render an estimate of its value and durability within the international division of labor. Although wage figures provided by animation company managers are impressively high (and unsubstantiated), they generally represent only the upper echelons—the highly skilled levels of the profession. Wages of the colorists, inkers, painters, cel cleaners, and even assistant animators—the individuals carrying out the drudge work, who far outnumber key animators and managers—are much lower. Of course, that is what attracts the foreign companies in the first place: large numbers of individuals willing to work hard for low wages in a stable setting.

Similar to other Third World workers in offshore electronics factories and data-processing backoffices, and reminiscent of pioneering cartoonists in the United States, Asian animators labor at tedious, eye-straining, and monotonous tasks for long periods, often forsaking weekend time off and annual vacations. Steady employment is not guaranteed because animation work is usually seasonal, and job security is almost nonexistent, especially for nonskilled employees. Just as U.S. animation companies farmed out jobs to Japan, South Korea, and Taiwan, so have those countries subcontracted to China, Thailand, the Philippines, Vietnam, or North Korea—a regional subdivision of labor—when their own labor became too expensive. The procedure has been to retain key animators and other skilled workers to perform pre- and postproduction jobs previously done in the United States, while transferring other work to havens of still cheaper labor.

As more domestic animation is produced in the television-rich countries of Asia, jobs at the lower levels are likely to open up again, although wages and working conditions are not expected to improve. In fact, they can be expected to remain substandard compared to those at foreign animation companies because locally owned companies usually are under less scrutiny.

Asian animators sound credible in their assessment that sufficient transfer of skills has taken place to launch and sustain domestic animation; certainly, thousands of Asians have learned parts of the animation business from the foreigners. However, they also have picked up other traits associated with such capitalistic endeavors such as valuing quantity over quality and commercialization over artistry. The famous Japanese animator, Hayao Miyazaki, commented on the commercialized nature of animated films in his country:

But regrettably, others making animated cartoons seem to be different. They insist, "This is the trend. . . . This is likely to be a hit. . . ." I can't imagine myself in such an inhuman task as making animated cartoons just to produce such things. To produce a decent animated cartoon requires anywhere from a year to a year and a half, and our private lives go out the window during this period. . . . Works of art are created by those who are prepared to go the limit. ("Money can't buy creativity," 1991, pp. 7-8)

One of China's premier animator/cartoonists, Zhan Tong, also decried the Western-oriented commercialization influences on his country's animators. He said many of them now concentrate solely on making money, and fewer are willing to devote their lives to artistic pursuits. Zhan confidently added that the whole situation will again be reversed (Zhan Tong, personal interview, August 15, 1993).

REFERENCES

Balnaves, N. (1996, January). International production and distribution. *Animation*, p. 123.

Cartoon news. (1995, September). *ASIFA San Francisco Newsletter*, p. 8.

Deneroff, H. (1994, October/November). Supply and demand. *Animation*, p. 18.

Edelstein, B. (1995, June 19-26). People still draw. *Variety*, p. 38.

Feazel, M., Galbraith, M., Demuth, N., & Gwynne, P. (1991, October). Asia gains from animation boom. *Asia-Pacific Broadcasting*, pp. 12-16.

Goldrich, R. (1982, August). Can American animators help losing work to the Far East?" *TV World*, p. 10.

Karp, J. (1995, June 22). Get it? *Far Eastern Economic Review*, pp. 88-89.

Klein, N. M. (1993). *7 minutes: The life and death of the American animated cartoon*. London: Verso.

Kurcfeld, M. (n.d.). Animation: What's next? New voices from Europe and Asia. *Animag*, 2(2), 31-36.

Made in Japan Disney cartoons. (1990, February 20). *Mainichi Daily News*, p. B-6.

Money can't buy creativity: Hayao Miyazaki. (1991, January). *Pacific Friend*, pp. 7-8.

Rodriguez, P. (1996, January). International production and distribution. *Animation*, p. 130.

Segers, F., & Michie, L. (1988, February 10). Yank coin to finance China animation site. *Variety*, pp. 1, 133.

Sirimas, C. (1993, October 14). Thailand's comic relief. *The Nation* (Bangkok), p. B-1.

Swain, B. (1995, June 19-25). Euro animation in I-market mode. *Variety*, p. 42.

Thornton, E. (1995, June 22). That's not all, folks. *Far Eastern Economic Review*, pp. 86, 88.

Walker, S. (1994, October/November). Strength through creative partnerships. *Animation*, pp. 50-51, 54.

Williams, M. (1991, June 3). Gallic toonster to ankle Asia. *Variety*, p. 43.

12

Restructuring Global Telecommunications: The Telephone Workers' Response in British Columbia

Sid Shniad and Charley Richardson

The unprecedented technological and political/regulatory changes that have transformed the world economy over the past 20 years have been captured by the term *globalization*. The concomitant restructuring of the world's telecommunications systems has led to an enormous increase in transnational corporate power and a regressive redistribution of wealth.

Promoted by powerful international institutions like the World Bank (World Bank, 1989), this new telecommunications model is designed to utilize the latest technology to generate economic efficiencies and organizational benefits for the transnational corporate sector. Meanwhile, the rest of society finds itself paying higher rates for inferior service, using outmoded technology.

Under an alternate model, similar technologies could be applied to satisfy all of society's needs for affordable access to the full range of information-based services that are becoming available ("A National Strategy," 1990). The approach ultimately used will be determined in the course of the struggle between the corporate sector and the rest of society.

This chapter looks at the origins of Canada's telephone industry—the role played by social regulation in the creation of that industry, the impact of U.S.-style deregulation on the system, the changes imposed by technology on the workplace, and the corporate blueprint for integrating its services and workforce into a more fully transnationalized telecommunications industry of the future. In addressing these issues, the chapter focuses on attempts by the Telecommunications Workers Union of British Columbia (TWU), the organization that represents telephone company employees in the country's westernmost province, to resist these regressive changes and to promote a more desirable alternative for themselves and the customers served by their employer, British Columbia Telephone Company (BC Tel).

THE ORIGINS OF TELEPHONE REGULATION IN CANADA[1]

BC Tel, which provides telephone service throughout the province, is a privately owned subsidiary of General Telephone and Electronics (GTE), a transnational communications giant headquartered in Stamford, CT. Until 1992, BC Tel—along with the country's other telephone companies—was a member of Telecom Canada, the organization that coordinated the activities of the industry, dividing toll revenues and ensuring that affordable services were provided throughout the country. With the approach of long-distance competition, the strictly administrative Telecom Canada was replaced by the Stentor organization, a consortium whose membership is composed of the incumbent telephone companies. Stentor has the mandate to coordinate their competitive strategies, including organization changes and marketing efforts. The formalization of the organizational change from Telecom Canada to Stentor marked a fundamental reorientation of the industry.

In June 1992 the Canadian Radio-television and Telecommunications Commission (CRTC), the federal regulatory body responsible for overseeing the country's telecommunications industry, approved the introduction of competition in long-distance voice service. Previously, Canadian telephone companies monopolized both local and long-distance service in their operating territories, while subject to a system of social regulation administered by the CRTC.

Under that regulated monopoly arrangement, Canada's telephone companies were required to provide universal service at

---

[1]This and the following section are drawn from "The Mulroney Trade Agreement" (1990).

affordable rates. In exchange for being allowed by government to control the highly profitable long-distance market, companies such as BC Tel were required to provide local service to every customer in their operating territory at prices that did not recover the related costs.

This use of long-distance profits to subsidize the price of local and residential service was the linchpin of Canada's socially regulated telephone system—a system that provided technically sophisticated services on a universal basis and at prices among the most affordable in the world. In the context of this de facto social accord, Canada's unionized telephone workers made a good living providing service to people in all walks of life and in every corner of the country. In recent years, the success of this system has been taken for granted.

To understand how the country's telecommunications system became so successful, it is important to examine the origins of Canadian telephone regulation. In 1880, Bell Canada, a subsidiary of the U.S.-based Bell Telephone, received a Dominion Charter from the Parliament of Canada, giving it the exclusive right to manufacture telephone equipment and to sell telephone service nationwide.

Bell Canada was cavalier in the exercise of its monopoly power, early on alienating residents in rural areas with minimal service. Return on investment in rural areas was simply insufficient to interest the company.

Over time, rural customers' resentment generated a public outcry, which led Parliament to withdraw the company's monopoly status. However, Bell continued to function as a de facto monopoly, using its market power to set rates that discriminated between business and residential customers, cities of different sizes, and rural and urban markets.

This de facto monopoly ended in 1893, when the patent owned by Bell's U.S. parent company expired. Independent (i.e., non-Bell) telephone companies sprang up, driving telephone rates down and dramatically shrinking Bell's revenues.

Moving to restore its monopoly position, Bell crushed the competing independents, denying them the right to hook up to its network and refusing to sell them equipment. These high-handed tactics generated tremendous public resentment, with the result that Canadians pressured government to end the company's maneuvers. They wanted the telephone business to be taken over and run as a government utility, along the lines of the postal and telegraph authorities in Europe. Aware of the imminent danger, Bell management promoted the idea that government should regulate the industry, rather than take it over.

GOVERNMENT-RUN PHONE COMPANIES:
THE ORIGINS OF CROSS-SUBSIDIZATION

By limiting its service offerings to areas where the rate of return was highest, Bell had neglected rural customers' phone needs. Some remote communities had no phone service at all. Tired of the low priority given them by Bell, residents of Canada's prairie provinces demanded that their governments establish provincially owned and operated telephone systems. The governments of Alberta and Manitoba responded first, setting up telephone systems in 1906, followed by Saskatchewan in 1908.

When Manitoba Government Telephones (MGT) was established, the province decided that all subscribers would be supplied service at affordable rates. Universal access was made a priority, even though service to remote and rural areas was provided at a financial loss.

To cover these costs, long-distance and business rates were set significantly above the cost to the province of providing them. The resultant surpluses were used to offset the losses incurred in rural and remote areas. This integrated approach to the provision of telephone service was called "cross-subsidization."

In marked contrast to the telephone systems owned and operated by provincial governments, Canada's private telephone companies were guided by the same bottom-line considerations that motivated Bell in the late 19th century. Universally available, affordable service was not a priority for them. Left to their own devices, these private companies would have continued to avoid rural and remote areas where their operations encountered high costs and insufficient rates of return.

As the country developed, however, the rise of new cities and towns created additional markets for telephone service. The way was paved for the creation of an implicit social contract: To avoid the prospect of being nationalized, Canada's privately owned telephone companies declared their willingness to accept a system of government regulation, under which they would provide service to residential and rural subscribers at a financial loss in exchange for monopoly control over highly profitable business and long-distance services in their operating territories.

In short, the country's private phone monopolies were anxious to capitalize on the burgeoning markets for long-distance, business, and urban residential telephone service and to counter increasing social pressure on behalf of universally accessible and affordable telephone service. They realized that the price they had to pay to maintain control of the lucrative side of the business was the acceptance of the concept of cross-subsidization that had originated in the government-run telephone companies on the prairies.[2] By having a monopoly, albeit a government-regulated one, these private telephone companies preserved their ability to overcharge business users and long-distance customers by a margin sufficient to pay for a generally socially beneficial regime.

TELEPHONE REGULATION UNDER ATTACK

Until the 1970s, the telephone industry in the United States was run under a system of social regulation similar to that of Canada. Unfortunately, the U.S. experience with telephone deregulation foretold developments in Canada.

Vincent Mosco (1989) explains:

> The trend to deregulation, which began in the 1970s and picked up speed in the 1980s, is essentially an effort to roll back this [regulated] system. . . . There are numerous reasons for this . . . suffice it to say that a change in the political balance sheet reflected the growing power of large business users who have fought for a telecommunications system that best serves their increasingly pan-national interests. Business came to realize more and more that the entrenched system of government regulation could not help them to reap the full benefit of new technologies. Although not entirely successful, large users have helped to dismantle a considerable amount of the national and international regulatory apparatus. These large users have reaped the benefits of their political pressure, not least because they are on the winning side of a massive redistribution of telephone rates. Over the past five years [Mosco was writing in 1989], long distance charges have dropped by about 35 percent, while local residential rates have increased by 46 percent or twice the rate of inflation. Individual customers, small businesses, and local organizations have subsidized the rate cut demands of large users. In essence, for a small group of big users, whose political clout has risen over the last two decades, deregulation and competition have meant . . . lower prices. For the majority of Americans, they have meant just the opposite. (pp. 32-33)

The business sector has always been hostile to government regulation of the telephone industry because such regulation mandated that long-distance voice services—the bulk of which is used by big companies—be priced above cost and that a significant portion of the resulting revenues be used to subsidize the cost of local service for the

---

[2]For a look at developments in the U.S. telephone industry, see Smythe (1970).

benefit of the general public. This arrangement was acceptable to the phone companies because they continued to enjoy a monopoly over the lucrative portion of the business and because the alternative—originally, at least—was a system of government-owned and -operated phone companies. However, government regulation and the resulting cross-subsidization was an ongoing source of irritation to the phone companies' corporate customers.

In a precursor to the regressive economic restructuring process known today as globalization, corporations in the United States, Britain, and Japan mounted enormous campaigns to force their governments to dismantle the social regulation of the telecommunications sector and to allow competition in the highly profitable long-distance part of the business. The companies' goal was to reduce their communications costs while reshaping the industry to more closely address their needs. As a result of these campaigns, the unitary telecommunications infrastructure, which had characterized the earliest phone systems in the United States, Britain, and Japan, was dismantled.

Contrary to what promoters of competition had promised, the results of these changes did not benefit ordinary telephone users. In fact, for all but the largest corporations, telephone competition had dramatic negative effects: New competitors invested huge sums in the duplication of existing facilities, and endless legal and regulatory disputes occurred concerning the financial terms by which competitors would be allowed to connect to each other's networks and by their adherence to these terms.

In the ensuing battle for market share, the quality of service provided the general public declined. Competing companies focused their organizational efforts on satisfying major business accounts, neglecting residential and rural customers. In the process, telephone companies reduced costs by laying off tens of thousands of telephone workers. With the end of cross-subsidization, local rates rose to cover related costs, causing residential subscribers to experience unprecedented increases in their monthly bills.

THE BATTLE AGAINST LONG-DISTANCE COMPETITION IN CANADA

Encouraged by success in the United States and backed by both Canadian and U.S. companies determined to lower their communications costs, would-be competitors and their backers put enormous pressure on the federal government to deregulate the country's telecommunications system and to open it to U.S.-style competition.

In 1984, CNCP Telecommunications, a partnership between the communications branches of Canadian National Railways and the Canadian Pacific conglomerate, applied to the federal regulatory body for permission to compete in the long-distance phone business. At that point, the Telecommunications Workers Union (TWU), which represents the unionized employees at BC Tel, decided to oppose the application. Aware of the U.S. experience with telephone deregulation and competition, the union believed that the interests of its members and those of the general public coincided in this situation: Competition and deregulation would lead to an increase in the price and deterioration in the quality of service, hurting both the workers whose jobs depended on providing that service and the customers who made use of it.

Given Canada's relatively small population, distributed over a larger area than the United States, the ratio of local and residential telephone costs to revenues is much higher north of the border. This creates a very large revenue shortfall in these services and makes the Canadian telephone system even more dependent on subsidies from monopoly long-distance revenues than the U.S. system had been. Consequently, Canada was potentially more vulnerable to the effects of competition and deregulation than the U.S. had been.

Building on this awareness, the TWU helped organize a coalition to defend the socially regulated system. Members of this coalition included unions, consumers, senior citizens' groups, antipoverty organizations, and the provincial governments that ran their own telephone systems. Uniting all of them was concern about the potential effects of a successful application by CNCP on the existing phone system.

Notably missing in the battle against the introduction of competition were Canada's private telephone companies. They saw the coming of long-distance competition as an opportunity to get out from under their obligation to serve rural areas and residential customers at below-cost rates and as a chance to lay off thousands of employees. Finally, they were confident of their ability to wage a successful battle against the potential competition. So instead of defending the merits of a socially regulated system, the phone companies stressed changes in the existing system that would place them in the strongest possible position when competition was introduced.

The social coalition defending the existing systems against competition and deregulation put up a tremendous fight, placing newspaper and television advertisements to alert the public about the negative ramifications of long-distance competition; producing information kits explaining the issue to potential allies; visiting federal, provincial, and municipal politicians to explain how competition would affect their constituents; and participating in the CRTC's regulatory hearings that examined the CNCP application.

During these hearings, the TWU cross-examined CNCP's witnesses and brought in its own expert witnesses to show the weakness of the company's case. Their evidence contradicted the rosy, ideologically charged picture that proponents of competition and deregulation had painted of the deregulated U.S. telephone system and shed light on the Canadian phone companies' self-serving position.

When the hearings were over, and the issue was in the hands of the regulators, the TWU helped organize one of the largest letter-writing campaigns in Canada's history. The goal was to convince the federal minister of communications, prime minister, and other politicians that it was not in their political interest to allow the CNCP application to be approved. The volume of mail was so large that the minister of communications, unnerved by its magnitude, instructed his staff to draft legislation that would have overturned the regulator's decision had the CNCP application been approved.

In the end, however, the government did not have to use this legislation. In a major victory for the union and its allies in 1985, the CRTC issued a decision denying CNCP's application.

ONCE MORE, WITH FEELING

Given the lure of the huge profits generated by long-distance telephone service, it was only a matter of time until another attempt was made to introduce competition in the Canadian telephone industry. In Spring 1990, Unitel filed an application seeking permission to compete in the Canadian long-distance market.

Unitel, a better financed, more powerful version of CNCP, was a partnership between corporate magnate Ted Rogers, owner of one of the country's two national cellular telephone companies and many of the largest cable television systems, and Canadian Pacific, one of Canada's largest conglomerates. Unitel enjoyed strong backing from both the Canadian corporate establishment, which wanted the lower long-distance rates that competition would bring, as well as the right-wing federal government in Ottawa.

CRTC's hearings on the Unitel application began in Spring 1991. Many of the same organizations that had opposed CNCP joined forces to oppose the Unitel bid, but the combination of Rogers's enormous financial resources, his allies in the corporate sector, and the procorporate orientation of Canada's right-wing federal government proved insurmountable.

Ignoring evidence about the regressive effects that long-distance competition had on ordinary telephone users in other countries, Rogers and his corporate and political backers emphasized that the governments of the United States, Japan, and Britain had already embraced telephone competition and deregulation. Enormous amounts of money were spent to convince Canadians that network competition was both inevitable and desirable. Despite the efforts of the union and its allies, the CRTC issued Decision 92-12 in June 1992, giving Unitel permission to compete in the long-distance telephone business.

This decision has been important for several reasons. It has meant that residential customers and those living in rural areas and small towns across Canada face significantly higher costs for their basic service. They are also at risk of being left without access to the latest telecommunications services. Finally, Canadian telephone workers are bearing the brunt of this decision as telephone companies restructure their operations to meet the needs of large corporate customers in major urban centers, in the process, shedding thousands of employees.

WORKING IN THE GLOBALIZED
TELECOMMUNICATIONS INDUSTRY

On the eve of divestiture [in 1984], AT&T was the world's largest private employer with over one million employees. . . . Since divestiture AT&T has eliminated some 140,000 bargaining unit jobs, while it has established and purchased major nonunion subsidiaries. . . . Since October 1993, major corporate restructuring accelerated. . . . US West announced the elimination of 9,400 jobs. . . . Bell South said it was eliminating 10,800 jobs . . . GTE announced the elimination of 17,000 jobs. . . . Pacific Telesis said it would downsize by 10,000 jobs at Pacific Bell. . . . AT&T declared it would eliminate another 15,000 jobs on top of already scheduled force reductions of 6,000 operator and call servicing positions and 7,500 jobs at Global Information Solutions, formerly NCR. . . . Ameritech said it would reduce its workforce by 6,000. . . . NYNEX . . . scaled back its plans to eliminate 22,500 jobs to 16,800 positions.

From the standpoint of labor-management relations, this massive industrial restructuring is in jeopardy of severing the traditional link between high productivity growth through rapid technological change and rising employee incomes with employment security. (Keefe & Boroff, 1994, pp. 1-5)

CHANGES, CHANGES EVERYWHERE

In these days of close partnership between government and business, we hear nothing but positive public relations messages about the wonders of the coming information superhighway. Setting aside for a moment whether the promises of a prosperous digital future will ever prove accurate, it is clearly the case that telecommunications workers—those at the front line in this sector—are experiencing massive change that is distinctly negative. Instead of being celebrated for their role in implementing the telecommunications revolution, they are truly its victims, threatened with becoming the "roadkill" on the information superhighway.

The impact on telephone workers is not limited to just job loss. The jobs that remain have been broken up and broken down, routinized, monitored, intensified, and deskilled. Highly skilled trouble-shooting jobs have been split into simple, repetitive clerical tasks. The work of operators, once requiring considerable knowledge of call routing, as well as the handling of complex time and charge schedules, has been similarly transformed into tiny tasks that are subjected to unimaginable speed up. Telephone workers, whose wages and working conditions were once the envy of the rest of the workforce, are experiencing a declining standard of living, increasing stress on the job, and an unprecedented level of insecurity about their employment future.

The information superhighway and the telecommunications industry of the future have been promoted as the key to the healthful recovery of the ailing economy. However, the cheerleaders for this scenario generally ignore the effect of the attendant changes on the industry's workforce. Even those who acknowledge that all is not rosy in the sector tend to dismiss problems as transitional and insignificant in the context of the huge benefits that will ostensibly flow to society when the full economic potential of the information superhighway is realized.

CHANGE IN THE INDUSTRY

Three intersecting trends are profoundly affecting the telecommunications industry, having an impact on the role of the workforce (and therefore the nature of jobs) and, in particular, undermining the ability of the workforce, through their unions, to have a significant voice in the future of work in the industry.

First, the very definition of the industry is changing. Not long ago, the telecommunications sector was known as the telephone industry. It was socially regulated and was defined by the transmission of voice service over copper wires. With the technological innovations of the past 15 years, however, telecommunications is now effectively deregulated and characterized by an ever-changing array of transmission media, including wireless, coaxial cable, and fiber optics, which carry an ever-expanding variety of data forms. The boundaries of this system, once readily defined, are now constantly changing.

The second major trend is toward an increasing and ever-changing array of players in the industry. Telecommunications used to consist, in essence, of a single regulated entity—the phone company—in each geographic region. Today, hundreds of different companies exist in every jurisdiction, each deploying an array of advanced technologies, offering new services, and actively engaging in attempts to undercut the traditional telephone company, as well as one another.

The third major trend consists of never-ending changes in the technologies that lie at the heart of the communications industry. Digitalization of the signal, computerization of administrative and switching functions, and the use of new transmission media—including fiber optic and wireless systems—all represent changes in core technology that act as enabling agents for the two trends discussed earlier, as well as for fundamental modifications in the way work is organized and in the conduct of labor-management relations.

The cumulative impact of these trends has been enormous. Telephone jobs were once characterized by a relatively high degree of job security, by work that was carried out at a reasonable pace, by wages and benefits that were relatively good, and by the existence of clearly defined job ladders and transfer rights, all of which practically guaranteed that workers' skills would be enhanced and that their income would grow over time. These attributes helped to create and maintain a stable workforce within the telephone companies. All of them have been undermined as the industry has undergone a series of fundamental changes.

The telephone companies' traditional willingness to pay decent wages, to provide relative employment stability, and to abide by the terms of an implicit social compact with their unionized workers was strengthened by several factors that gave telephone workers significant bargaining power with management. First, the technology at the heart of the telephone system was far from dependable. Because the system was highly sensitive to any problems that arose, these had to be addressed immediately. The failure of a transmission line or a switching frame could easily disrupt service to a number of customers. As a result, telephone companies needed large numbers of specially trained and skilled workers to keep the system functioning smoothly.

Second, the skills needed to keep the system operating were often unique to the telephone system and were normally acquired through on-the-job experience at the phone company, rather than through technical training acquired outside. These skills were essential to the maintenance of switching equipment and transmission lines, the handling of billing activity, and interaction with the customers, but tended to be of limited usefulness in other industries. That these workers and their skills were not easily replaceable served to increase the bargaining power of the workforce. At the same time, the limited applicability of telephone workers' skills tended to tie the workforce to the company. In other words, there was a relationship of reciprocal dependence.

Third, the absence of competition gave phone companies flexibility over the pricing of their services and the ability to carry out long-term planning. These factors, coupled with a constant, incremental increase in productivity within the industry, allowed companies to meet the wage increases that were necessary to maintain a stable workforce. At the same time, the fact that they could engage in long-term planning gave phone companies the ability to train their workforces as an integral part of implementing their capital investment schedules. That is, telephone companies were able to coordinate the introduction of new technologies with their human resource planning and long-term financial considerations. All of this, in turn, reinforced the terms of the implicit social compact that existed within the industry.

FUNDAMENTAL TECHNOLOGICAL CHANGE

Fundamental technological changes have both pressured and allowed the telephone companies to abandon the historical social compact. Current telecommunications technologies are designed to increase the operational efficiency of the system and to make it fault-tolerant. Both of these factors lessen management's dependence on skilled employees. AT&T, for example, brags in its commercials that there are now 134 alternative ways to route a call should there be trouble in any one of its switching or transmission systems. Thus, at the same time that competition is undermining companies' control over pricing and their ability to carry out long-term planning, new technologies are increasing the reliability of internal systems. This allows companies to handle a growing volume of traffic with a shrinking workforce, a fact which has had a particularly powerful impact on the unionization levels in an industry that was once entirely unionized.

A vicious spiral has been set in motion. Telephone companies, functioning in today's deregulated environment, are pressed by stockholders and capital markets to use technology to expand their service offerings and by competitors to cut their prices, particularly in the lucrative business services market. The strategy that companies use to keep ahead of the competitive pack centers on the increasingly rapid introduction of new technology, a radical downsizing of the workforce, and entry into entirely new markets. Taken together, these moves have put enormous strain on the longstanding social compact that has existed between companies and their employees.

Intensifying the impact of these influences is the fact that startup companies entering the telecommunications industry today have never been party to any kind of social compact with their employees. At the very time that new technological advances are making it possible for incumbent companies to break the longstanding compact, new competitors are using the same new technology and operating in an environment free from formal, legal regulation by the government or from de facto social regulation in the form of social compact or union contract. This, of course, creates further pressure on the traditional telephone companies to free themselves of the social constraints imposed by unions.

The effects on the unionized workforce have been devastating, as cited, for example, in Keefe's and Boroff's (1994) statistics on job losses in the U.S. industry. Moreover, the jobs that have survived are significantly less unionized because of higher end technological skill transfers outside the bargaining units and the contracting out of much of the remaining work to nonunion firms. To make matters worse, the unionization rate is less than 1% in the new wireless portion of the industry. When we examine today's wireless and telephone industries as a single sector, it is clear that the overall unionization rate is dramatically lower than it was in the traditional telephone industry.

The changing definition of the industry has also worked to undercut wages. Bell Atlantic technicians are paid US$18.95 per hour, with a benefits package that includes health and pension plans. In contrast, technicians at TCI, one of the largest cable operators in the United States, are paid less than $10 per hour and have no health insurance or pension plan.

When the cable and telephone businesses were separate industries, telephone companies could tolerate the payment of significantly higher wages. As a result of the convergence of their respective technologies, however, the two sectors are in the process of becoming a single entity in which cable and phone companies compete for the same customers. A wage differential of this magnitude cannot survive under such circumstances.

Whereas workers in the telephone industry used to enjoy a relative degree of job stability, the lives of today's telecommunications workers are characterized by constant change and increasing job insecurity. A large proportion of the work has disappeared as computers have taken over the functions of monitoring the system, diagnosing troubles, receiving customer requests for changes in service, administering the system, assigning costs, doing billing, and carrying out collections.

Unprecedented job losses for some is accompanied by unprecedented stress for those who survive the never-ending downsizing that is wracking the industry. A combination of relentless automation, unending reorganization of work, and the relocation of remaining jobs to new localities has transformed what was once a stable work environment into one characterized by unremitting change.

Robotic operators that recognize the human voice, for example, are being used to eliminate thousands of jobs, and the work that remains is greatly intensified. Contrary to popular opinion, however, automation is not being applied exclusively to "unskilled" jobs. The "expert systems" that are being developed and implemented are capable of handling the vast majority of switching problems, leaving companies with a need for relatively few super-skilled technicians. In addition, these top-end jobs are increasingly being done by managers, instead of the unionized workforce, thus reinforcing the deunionization trend discussed earlier.

Large numbers of repair jobs that once required highly developed skills are being transformed into a series of clerical tasks that involve the punching of customer information into highly sophisticated computer systems whose software has the necessary technical information built into it. Running diagnostics, which used to be a highly paid skilled job, has become a series of mundane tasks based on following decision-tree instructions that explain how to replace defective circuit boards.

In the context of this never-ending change, telecommunications companies have downgraded their traditional commitment to train the workforce internally. Although a few upgraded jobs now receive advanced training in computer science and electrical engineering, management is increasingly able and willing to hire people with the requisite skills off the street.

The remnants of the internal labor markets in the old telephone companies have lessened the threat of layoff for many, but at a high personal cost. Many surviving telephone workers have been forced to choose between their jobs and their communities. The Atlanta service megacenter, which handles AT&T's switching operations for most of the

East Coast of the United States, is populated by telephone workers who originally worked at locations all across the country. This centralization of operations is made possible by the same new technologies that allow management to constantly reorganize work and to downsize the workforce.

Centralization also helps to undermine the social compact between telephone companies and employees, as it tends to undermine the cohesiveness of the unions. Unions are made up of social groupings, based on people who work together. These are profoundly disrupted by managerial decisions that transfer employees from one geographic location to another on an ongoing basis.

There is nothing inherent in the technology that requires it to be used to centralize work. The technology only makes such centralization possible. If companies were taking their employees' interests into account, they could use the same technological advances to decentralize their operations and to situate them in the communities where their employees live.

Wave after wave of unprecedented change, combined with the destruction of the social compact, means a loss of security even for those who still have jobs. Employees whose job future is fraught with uncertainty are nevertheless forced to become acquainted with new versions of software, new product and service offerings, new equipment, and new workstation environments. Workers are insecure because they fear they will lose their jobs, and these fears are compounded by the fear that even if they are not displaced, they will be unable to meet the demands of ever-changing and ever more stressful jobs. It is not mysterious that telephone workers, especially operators and commercial marketers, are reporting an enormous increase in job-related anxiety.

MONITORING AND SPEEDUP

New technology has given telecommunications companies the ability to monitor their workforce to an unprecedented degree. Monitoring is usually discussed in connection with operators, whose "average work time" is measured constantly. However, monitoring is increasingly applied to technicians and others, including the once relatively autonomous installation and repair workers who make service calls in phone company trucks. Management's use of monitoring, and the accompanying pressure to speedup, are also major contributors to the stress-related disability and repetitive strain injuries that have become epidemic in the industry.

THE HIGH SKILL, HIGH WAGE MYTH

Telecommunications is often described as an exciting and dynamic, technology-based sector characterized by highly skilled workers who enjoy well-paid jobs. It is clear from the previous discussion that the benefits being generated in the sector are translating into bonanzas for company owners and disasters for their employees. Yet telephone companies are spending large amounts of time and money to promote elaborate "employee involvement" programs. If, as we are arguing, management is holding all the cards, why is it going to such lengths to win the hearts and minds of its employees?

Telephone companies are simultaneously augmenting their capital spending and reducing their operating costs. High levels of investment are needed to respond to niche markets and to keep ahead of the competition, both in providing new services and in reducing costs. Although this investment eliminates jobs and further reduces management's reliance on what remains of the traditional workforce— particularly those segments that are unionized or in possession of noncertifiable skills—the effective use of the new technologies depends on the cooperation of particular segments of the workforce. As a result, many companies—particularly the unionized ones—have separated the "high value-added" parts of their businesses, in which workforce morale, worker involvement, and the alignment of workers' and managers' priorities are essential ingredients for success, from the parts of the business in which little value is added and in which cost-cutting, downsizing, increased managerial control, relentless elimination of jobs, and speedup of work are the order of the day.

CORPORATE IMAGES, LABOR REALITIES

Business units' different strategies reflect these conflicting priorities. Thus, although some parts of the industry stress the need for teams and mutual gains, others emphasize monitoring, speedup, and ever-greater doses of automation. Everywhere we look today, we see evidence of corporate public relations campaigns that are selling the image of an emerging globalized information industry bringing us a wonderful world of useful services, provided by happy workers with interesting, highly skilled, and well-paid jobs. Unfortunately, the reality that workers in today's telecommunications industry are experiencing bears no resemblance to this advertising agency fantasy.

If the transformation that is taking place in the telecommunications industry is to yield benefits for the people who depend on it for their employment, the corporate strategies that are driving these changes must be challenged.

DEREGULATION AND COMPETITION:
THE COMPANY'S RESPONSE

In the aftermath of the CRTC decision giving the go-ahead to long-distance competition, BC Tel announced a "strategic renewal," the purpose of which was to restructure the company and to improve its ability to function in the new, competitive environment. Expensive consultants were hired and, after a lengthy and costly study, BC Tel came forth with the Gemini Project, designed to reorganize work within the company. This internal reorganization is part of a restructuring effort taking place across the Canadian telecommunications industry. Even before the 1992 decision, the country's telephone companies anticipated changes, replacing Telecom Canada with Stentor, whose purpose was to more closely coordinate the activities of its member companies.

Stentor immediately became a member of the Financial Network Association (FNA), an international consortium headquartered in Brussels with the purpose of meeting the international telecommunications needs of banks and financial companies. FNA's member companies include U.S.-based MCI International, France Telecom, Deutsche Bundespost Telekom, Hong Kong Telecom, KDD of Japan, RTT-Belgacom of Belgium, Italcable of Italy, Singapore Telecom, Telefonica of Spain, Mercury Communications of Britain, and AOTC of Australia. FNA is designed to allow a bank to make the telecommunications service arrangements needed for its foreign banking operations by dealing with a domestic phone company.

Like other telecommunications companies of North America, western Europe, and elsewhere that function under deregulation, BC Tel focuses on satisfying the needs of high-profit, big-business customers located in major urban centers. The company, following the lead of the Stentor organization, is emulating a strategy pioneered by Ameritech, one of the Regional Bell Operating Companies (RBOCs) that was spun off when AT&T underwent divestiture in 1984. The RBOCs were given responsibility for local and short-haul long-distance service within their operating areas.

If followed in Canada, Ameritech's plan could have far-reaching consequences for the country's telecommunications system. Ameritech has deaveraged its rates and introduced Local Measured Service in some states. De-averaging means that people living in thinly populated and remote areas will be hit hard as rates move closer to their underlying costs. (The flip side is that phone companies will drop their rates in more heavily populated urban areas, where their competitors are focusing their efforts.)

As well, Ameritech has moved to "incentive-based" regulation, whereby phone companies increase profits by reducing their costs—particularly by cutting staff. This regulatory change has been the basis for the loss of tens of thousands of jobs in the deregulated U.S. telephone industry. Ameritech is also proposing to open local exchanges to competition as a tradeoff for getting permission to enter the long-distance and cable TV markets and sharply reduced regulation of competitive local markets.

A key part of Ameritech's proposal is the idea that financial contributions toward subsidies will be paid by all companies, not just the dominant carriers. This would unite all companies in the battle to dump subsidies and, if successful, would result in reduced overall demand for communications services. While such a move would be good for the phone companies, it would be bad for phone workers and the general public. According to Ameritech, the "drivers" behind this proposal include: (a) competition, which is accelerating loss of market share in profitable market segments; (b) companies' need to speed up their rate of depreciation in order to write off their investments more rapidly, thus reducing the amount of taxes they pay and increasing their after-tax profits; and (c) the existing system of price averaging, which keeps phone companies from dropping their rates on the high traffic routes, thus allowing their competitors to focus on these routes with target marketing.

While these changes are being considered, there is taking place a rapid convergence between the communications and information industries. Since Ameritech proposed its strategy, the RBOC known as US West struck a deal with Time-Warner, MCI formed a partnership with 150 cable operators and wireless service providers, British Telecom and MCI formed an alliance, and AT&T acquired the McCaw Cable company. The question is: What will be the effects on ordinary phone customers and telephone workers as the underlying corporate strategy unfolds?

Ameritech's goal is to create a "new market segment," referred to as "the communications intensive household," which represents the top 20% of the company's consumer base, generating about 70% of its long-distance revenues. It is this market that is being targeted for second lines, voice mailboxes, and special central office features. Under the Ameritech plan, the remaining 80% of the customer base will receive significantly less attention.

The degradation of phone service is not confined to the areas of the business that are under competitive pressure. In those areas where BC Tel still has a monopoly position, it is trying to increase its rates. Under an application filed with the CRTC in early 1993, the company sought to increase average monthly local rates for individual-line residential service by 35% and to increase the cost of business service by 10%. BC Tel also has received permission to charge customers for directory assistance and to abandon its responsibility for installing inside wire on customers' premises. The latter is a money-losing service that the company has offered as an integral part of its regulated monopoly service since its inception. These moves make sense from the phone company's perspective, but they do not bode well for phone workers or the general public.

In June 1993, when the company's application for an "interim" local rate increase was denied, BC Tel laid off 800 temporary employees. This was shortly after the company announced the seventh successive annual increase in common share dividends, which, according to BC Tel's president and CEO, "reflects confidence in the Company's long-term earning potential" (BC Telecom, 1993).

In short, under the new, globalized, restructured regime emerging in the telephone industry, corporate priorities are in stark conflict with the needs of telephone users and telephone workers. Ignoring the company's continuing profitability, management demands enormous productivity gains despite the fact that productivity increases in the Canadian telephone industry have significantly outstripped those in the rest of the economy for the past five years.

Despite threats of layoffs and the use of sophisticated electronic equipment to monitor employees' performance (an integral part of the company's continuing emphasis on top-down control), managers see no irony in using a language of "empowering" employees and getting them involved in a "team" approach to providing better service. Compounding the irony is the fact that employees know areas of the company in which management has tolerated violations of regulatory quality standards for years without taking corrective action. BC Tel continues to insist that its reorganization plans are essential to remain competitive in a changing world.

To the union's pleasant surprise, however, the threat posed by external competition has made the company very solicitous of the union's views. Management is clearly concerned that the TWU could become an obstacle to the changes that BC Tel deems necessary in face of the new competition.

This puts the TWU in an ambiguous position. Union members recognize that the company must respond to competitive pressures.

However, they have concerns about a strategy protecting only the most highly profitable accounts while neglecting whole classes of customers and entire regions of the province. The union has no objection to improvement of BC Tel's financial position, as long as it is by expanding services and revenues—and thereby employment—rather than by cutting costs through a reduction in service offerings.

BC Tel is emulating the Ameritech strategy, which gives high-end customers everything they want while neglecting the needs of the majority of phone users. As an alternative, the TWU has promoted what it calls service enhancement, a strategy designed to satisfy the communications needs of all customers.

The central idea in the union's approach is to expand the provision of high-quality, state-of-the-art telecommunications service to tens of thousands of residential customers, community organizations, and small businesses across the province—users who could be left without access to these services if BC Tel pursues the Ameritech strategy. This alternate strategy offers union members the prospect of employment based on plugging customers into a range of expanding informatics networks and then showing them how to use the services for personal and community welfare.

In the view of the union, the company's responsibility to the people of British Columbia can best be fulfilled if BC Tel would adopt a strategy that expands the use of telecommunications to link people in all walks of life and in all parts of the province, thereby providing affordable access to information-based services, instead of restricting the deployment of resources to meeting only the needs of large companies located in major urban centers. If successful, the union's approach will generate high-quality, well-paid telecommunications jobs, which will help the public increase its use of new, information-based services. This approach will mean that instead of shedding employees, the phone company will have to hire and train people to do the high performance work necessary to provide the full range of sophisticated telecommunications services.

In short, instead of an organizational model based on tight managerial control of a shrinking number of employees, this counterstrategy will require that a growing number of knowledgeable individuals—TWU members—have the skills, training, and understanding necessary to execute a strategy based on service enhancement.

CONCLUSION

In the brave new globalized world, workers are being told that labor-management conflict must be consigned to the history books. But the TWU views the situation differently. Although the interests of the TWU and those of the British Columbia Telephone Company are linked, these interests are far from identical. An emphasis on management's unilateral right to run the company or to engage in mindless cost cutting—even when these are carried out in the name of globalization—hurts both TWU members and their customers and compels the union to resist management's misguided priorities.

The union does not want to put BC Tel out of business, but it does realize that in the age of globalization, it must play an active role in the restructuring process that is transforming the world of telecommunications in order to protect the interests of its members and the general public.

REFERENCES

BC Telecom. (1993, July). *2nd Quarter Interim Report.*

Keefe, J., & Boroff, K. (1994, June). *Telecommunications labor-management relations one decade after the AT&T divestiture.* Paper presented at the conference "International Developments in Workplace Innovation: Implications for Canadian Competitiveness." Toronto, Canada.

Mosco, V. (1989, July/August). "Deja vu all over again?" *Society*, pp. 31-38.

The Mulroney Trade Agreement—a threat to Canada's telephone system? (1988, January). Unpublished paper. Vancouver, British Columbia: Telecommunications Workers Union.

A national strategy for the information age: The Telecommunications Workers Union response to the Department of Communications' request for submissions on Notice No. DGTP-09-89 regarding Local Distribution Telecommunications Networks. (May, 1990). Unpublished report. Vancouver, British Columbia.

Smythe, D. (1970, June). *The relevance of United States legislative-regulatory experience to the Canadian telecommunications situation, a study for the Canadian Department of Communications.* Government of Canada.

World Bank—The International Bank for Reconstruction and Development. (1989). *Restructuring and managing the telecommunications sector, A World Bank symposium.* Washington, DC: Author.

13

Trade Union Telematics for International Collective Bargaining

Dave Spooner

The international trade union movement is increasingly equipped with access to tools of information technology (IT) and telecommunications. The study of the use of electronic mail by Celia Mather and Ben Lowe (1990) showed that internationally at least 180 labor organizations were "online" in 1989-1990. In 1990, the International Confederation of Free Trade Unions (ICFTU) commented that the value of electronic communications was "clearly established" ("Towards effective responses," 1990), as increasing numbers of international, national, and local unions are, or plan to be, online.

In effect, large sections of the trade union movement can operate as an electronic global network. However, it is considerably more complicated to develop a global information system that has genuine practical value. Although much remains to be done simply to assist trade unions to go online, there is a growing need to consider a longer term information technology strategy—particularly in relation to collective bargaining within transnational corporations.

INDUSTRIAL CONTEXT OF INTERNATIONAL BARGAINING
AND TELEMATICS

In 1983-1984, a number of trade union researchers tried some early experiments to assess the potential value of electronic communications for unions as part of the infant Popular Telematics Project (POPTEL). This involved coordination with workplace representatives from Unilever plants in London and the Netherlands, and Ford workers in London, linked to the work of the Greater London Council to assist workers in transnational corporations (TNCs). Apart from crude messaging between the United Kingdom and the Netherlands ("Hello, is there anyone there?" etc.), the project achieved no practical result, although we learned a lot. We became cognizant of the huge differences between the nature of the company itself, its style of local and international management, the problems thus faced by unions attempting international cooperation on bargaining issues, and the very different information technology strategies needed.

The examples of Unilever and Ford are instructive. Unilever covered a vast spread of product lines employing workers in diverse areas such as soap making, edible oil refining, farming, plantations, transport, and food production. The conglomerate operated through a large number of semiautonomous companies, some of which were household names (Birds Eye, Walls, Lever Brothers, etc.); indeed, many Unilever workers were unaware they were employed by a giant transnational corporation. The power of Unilever was primarily exercised through finance and a permanent process of merger, acquisition, and disposal of subsidiary companies, with local management maintaining a relatively high degree of autonomy over industrial relations policies and management style, despite Unilever's attempts at developing a corporate management culture. Most collective bargaining, in the U.K. at least, was conducted at a plant or site level, with relatively high levels of unionization and local collective bargaining agreements.

The unions concerned (particularly the International Union of Foodworkers [IUF] and its affiliates) faced the need for high-grade intelligence of Unilever management strategies, primarily to assist local bargaining and to strengthen their arguments over transborder negotiating structures.

There was, therefore, little perceived need for the development of workplace-to-workplace electronic links, for what would a fish factory worker in northeast England have to talk about with workers in a Dutch oil refinery or on a West African plantation? There were possibilities (as pioneered by the IUF around cocoa production) for pulling together chains and subchains of production in and around Unilever, but this seemed to have few short-term gains for the coordination of union strategies for Unilever as a whole.

What was needed desperately was to pool, analyze, and disseminate intelligence about the company at a corporate level, which could be greatly assisted by the development of networks between national and international union researchers and negotiators, access to commercial databases, electronic access to experienced database users ("database brokers"), databases of trade union research materials, and so on.

The Ford Motor Company, however, had a relatively restricted range of products and, unlike the situation at Unilever, the skills and production processes required of one local or national workforce were directly comparable with those in other plants. Ford's corporate power rested on its integrated global production, and, indeed, the company positively encouraged its workforce to think in global terms—pitting plant against plant and country against country in negotiations over wages and productivity. Collective bargaining was conducted at the plant or national level, although key management investment, production, and industrial relations strategies were clearly decided at a world-regional or global level.

The IT implications and potential for unions within the auto industry, therefore, looked very different from those of companies such as Unilever. Less was to be gained from a major investment in the use of commercial databases as the information they contained was normally of a lower quality than that which was common knowledge on the Ford shopfloor. The greatest need was for an international coordination of information, strategies, experiences, and policies between union negotiators and representatives and for that information to be collated, analyzed, and fed into union policy making at an international level. This would generate a need for communications tools and networks (whether through e-mail, fax, telex, or face-to-face meetings), overcoming problems of language difference, cost, and so on, rather than for purely informational tools such as commercial online databases.

From these examples, it is readily apparent that the IT and telecommunications strategies developed by different trade unions must reflect the very different collective bargaining contexts within which they operate. This has already has already begun to happen, as a comparison of the use of telematics by the International Chemical and Energy Workers Federation (ICEF) and the International Transport Workers Federation (ITF) shows (see ICEF, 1990).

University of Warwick researcher, Paul Marginson (1982), who has looked at questions of transborder collective bargaining within Europe, identifies three factors that influence management strategies on collective bargaining: the historical roots of particular TNCs, the management culture within the TNC, and its diversification strategy. According to Marginson, where operations in different countries have been established on new "greenfield" sites, the company is more likely to have a common international approach to industrial relations and an internal management structure capable of handling transborder trade union negotiation (whether or not it chooses to do so). In these circumstances, industrial relations are more likely to be comparable from plant to plant across national boundaries; hence, there is considerable potential for electronic site-to-site trade union communications to coordinate industrial information.

However, companies that have expanded internationally through a process of merger and acquisition are more likely to have a wide range of management styles and industrial relations customs and practices inherited from older established practices, from which it is far harder for management and unions to develop coherent transborder industrial relations agreements. Even if they are involved in comparable production processes from site to site, the potential for trade union electronic coordination of collective bargaining is considerably reduced.

Closely related is the distinction between "geocentric" companies, with a highly international management structure transcending local cultures and boundaries, and "polycentric" ones, involving local participation in ownership and locally determined management. As Charles Levinson (1972) has argued, geocentric companies are most vulnerable to trade union demands for transnational collective bargaining. They also have a greater potential for the development of computer-based trade union communication networks and common international databases of trade union information, derived and accessed from site-based union representatives.

Useful distinctions also can be made between companies or industries according to their strategies for diversification. Those such as Ford, which concentrate on economies of scale, have a highly developed, vertical integration across international frontiers and require centralized production scheduling and tight global coordination. This makes the entire operation vulnerable to local trade union action. When there is coordination by international company councils or similar structures, cheap and effective computer communications can be highly useful.

Others companies such as Unilever, based on economies of scope, are primarily organized through horizontal integration, dependent on shared marketing, distribution, and technology strategies in the various subsidiary companies. They require strong interdivisional coordination (normally through financial control) by the corporate headquarters. The ability of trade unions to influence the global operation of the company through local industrial action is extremely limited, although considerable potential exists for corporate campaigns (such as ethical investment/disinvestment strategies) that attempt to affect the company's marketing image or political standing. (There have been some effective partnerships developed by international campaigning groups and trade unions in alliance with human rights or environmental organizations.) Computer communications has already shown its tremendous value in providing a vehicle for such coordination between trade unions and "social movement" organizations or nongovernmental organizations (NGOs).

In essence, for unions dealing with geocentric, greenfield-based companies with strong vertical integration, an international telematics strategy could include site-to-site electronic links, databases of trade union-derived information based on direct site-to-site comparability, international networks of local negotiators, and so on—effectively, an information technology strategy designed to empower transnational collective bargaining based on local strength and compatibility.

However, for those unions dealing with polycentric, acquired/merged companies with strong horizontal integration, a likely telematics strategy would concentrate on national and international centralization of company intelligence, the training of trade unionists to use commercial databases, strong telematics links between trade unions and NGOs, and durable electronics communications between national negotiators and researchers—effectively, an international communications and information system designed to support locally-based collective bargaining.

There is an outstanding and exceptional model among companies that have a truly international service delivery—notably transportation companies, hence, the unique nature of the International Transport Workers Federation's IT and telematics development.

Therefore, if trade union telematics develops to its true potential, a complex and shifting web of communications networks and information sources will emerge, encompassing established national and trade union structures with strong vertical lines of communication; new horizontal networks running from national union to national union, research department to research department, workplace to workplace and from NGO to union; and more ephemeral and "organic" structures established to deal with specific events or campaigns that cross all paths.

Above all, this requires flexibility—both in terms of the choice of technology and resource management. Ironically, it is almost a mirror image of the strategies adopted by many employers, as they come to grips with the full potential of telematics for international production and management.

TECHNICAL CONSTRAINTS

Major technical constraints stand in the way of developing a comprehensive worldwide telematics strategy for the labor movement internationally. These include cost, lack of network access in many countries, and problems of training and awareness (sometimes called "acculturation"). "Acculturation" includes developing a knowledge of the technology's capability, who is online elsewhere, with whom communication is valuable (and, indeed, who should be avoided), the most important and relevant sources of online information, and so on. Many of these problems are addressed elsewhere.

Language differences are a particularly thorny problem, which telematics will not be able to solve for a few decades. Language transmission is considerably more a problem for those working in non-Romanic scripts, as in Asia, the Middle East, and elsewhere, than in Romanic scripts. Various organizations sympathetic to the labor movement have made considerable progress in international communications using Chinese or Arabic. Even so, most Asian trade unions will continue to rely on fax technology for a long time.

Language differences coupled with the lack of telecommunications infrastructure mean that for the next decade or more, international trade union telematics must continue to rely on a mix of electronic mail, fax, telex, and airmail.

A tremendous danger exists of developing "two worlds" of international trade unionism, one with access to fast and efficient communications and information, and the other reliant on traditional media. If not addressed, this split will have profound effects on the processes of democratic participation in the international movement.

POLITICAL CONSTRAINTS

Political divisions and rivalry between trade unions remain a fact of life, even though the recent collapse of World Federation of Trade Unions (WFTU) structures has dramatically improved the broad international political climate. Understandably, many national unions are still extremely reluctant to cede authority to international trade union structures and more still are uncertain about, or opposed to, workplace-to-workplace transnational contact.

If the application of telematics is to develop as a serious tool for transnational collective bargaining, two key political issues must be addressed: the ability of unions at workplace level to use the technology (particularly important for unions facing TNCs of the "geocentric" model), and the pooling of trade union intelligence and information into accessible online databases on an international level (particularly important for unions facing "polycentric" TNC employers).

There is no doubt that the technology now available provides, at least in theory, a major challenge to established patterns and practices of trade union organization, particularly at national and international levels. For a local union to be online, it is technically as simple, fast, and cheap to send direct messages to dozens of unions across the globe as it is to send one message to a single union in the next town and, in theory, to gain access to the same information, debates, and news as any national or international trade union federation. On the minus side, this could undermine the authority of some national and international trade union structures and diminish their role in strategic collective bargaining.

However, even after a number of years of telematics development, the much feared bypassing of national and international trade union structures in favor of "horizontal networking" between local union committees has, by and large, failed to materialize. Although impossible to prove given the confidential nature of most electronic mail, most "horizontal" communication is probably informal contacts between researchers and organizers within national unions, international unions, and labor movement-related NGOs. Electronic communication between unions themselves remains largely mediated through established formal structures, both national and international. It will almost certainly remain that way, as long as affiliated organizations regard the political and strategic roles of national/international structures as relevant and valuable.

Trade union telematics (if it continues to develop) will, however, encourage a qualitative shift in this relationship and provide both new opportunities and threats to traditional trade union structures.

The facilities are now in place for the trade union movement to develop its own sophisticated, full-text, online databases to a high professional standard, with global access and low cost. This is an extremely important development, potentially adding an entirely new dimension to the trade union telematics framework.

Technically, the construction of such a database host for the labor movement is extraordinarily simple, only requiring a potential database provider to send disks of raw text of newsletters, policy documents, research papers, and so on to the host system, where it can be converted into a highly powerful, fully indexed database at the touch of a few buttons. Politically, however, it raises some old problems in stark relief. Industrial relations academic Tony Lane (1987), in a controversial (some would say infamous) article, argued that trade unions were ill equipped to cope with modern demands for industrial democracy. He said that the rate of exchange of economic intelligence between trade unions was almost zero, further stating it made no sense to talk of a community of trade union researchers. They had, he argued, "no natural arena for meeting, no formal machinery for exchanges, and talk of a common database would bring ceilings down with ribald, cynical and incredulous laughter" (Lane, 1987, p. 21).

Nearly a decade later, considerably more awareness of the need for such interunion cooperation exists, and the laughter might be a little more restrained; yet, a serious problem certainly persists: how to organize a common, pooled database system that contains information and analysis of sufficient value to assist collective bargaining to a serious extent, while enabling national and international trade unions to retain control over rights of access.

To a certain extent, the technology itself can be "tweaked" to provide safeguards. Individual databases can be structured to allow only affiliated member access, to charge nonmembers a higher rate, or to grant access to specifically designated password holders. This severely limits, however, the opportunities. The potentially enormous value of a shared database system would be virtually eliminated if it were to degenerate into small, discrete, and exclusive islands of information, and it would be unfathomable to all but research professionals.

Yet there is a logic to the reticence of trade unions to put valuable information and intelligence on collective bargaining issues into a common pot (apart from the obvious fear of access by employers). The reality is that in the U.K., at least, the unions are fighting a battle for their very survival in the face of falling membership and political attack. The quality of assistance given to members' collective bargaining efforts by their national union is one crucial factor that could determine whether memberships rise or fall. Also, unions' support and attractiveness to other unions becomes an issue when they negotiate mergers, inevitably invoking interunion competition. The strength and reputation of the union's industrial research and intelligence capabilities—its knowledge base—is, therefore, a powerful asset in these matters, thus, the understandable reluctance to "give away" high-grade information.

Nevertheless, trade unions have joined forces to pool important intelligence, a case in point being public sector unions in the U.K. sharing databases on privatization. On a broader stage, the capacity for democratic trade unionism to gain a foothold in transnational bargaining would be radically transformed by the establishment of a well-organized global information system. We are very close to seeing, for example, a trade union negotiator from virtually anywhere in the world being able to enter a database system, type in the name of a company, a hazardous chemical, or a industrial relations research topic, and, at a very low cost, being able to retrieve background information, policies, or bargaining data from national or international trade union sources worldwide.

ELEMENTS OF A STRATEGY

It is now possible to piece together elements of an international trade union strategy for the development of telematics, not for internal trade union administration, as important as that is, but for national and international collective bargaining. So far, no single trade union application of telematics for collective bargaining exists; a variety of communications technologies must be used, depending on the industrial context.

Trade union "information audits" may be required to assess the intelligence needs and information resources related to particular companies and industrial sectors, identify the major sources of information required for effective bargaining, analyze key information flows within and between the organizations, and assess the likely training needs of national and local representatives as appropriate. Such an analysis (not necessarily as grand as it sounds) then needs to be placed side by side with the telematics tools available, and a detailed "bespoke" telematics plan developed.

Although this is blindingly obvious, many trade unions still have developed information technology strategies as a poor relation or adjunct to the use of computers for internal administration, record keeping, and accounts, and they have never considered it in the context of collective bargaining needs, research capability, or fundamental industrial organization. Urgent attention has to be paid to the financial and technological divisions between North and South, particularly to offset the danger of European unions simply concentrating on inter-European networks and information banks.

Simple financial support is needed for unions that face potentially crippling costs of hardware and international communication. Some unions have already been the recipients of aid from international unions and development agencies.

When direct computer communication proves impossible, whether because there is no adequate telecommunications infrastructure or because of language barriers, steps are needed to guarantee that affected trade unions are not left isolated. In practice, this may mean that organizations are given the responsibility for taking important documents or discussions from the networks or searching databases and passing the information on by more conventional means.

This already has been achieved with national or regional organizations acting as brokers or "relay stations" for local trade unions. For example, Asia Monitor Resource Centre, based in Hong Kong, received faxed and airmailed information and solidarity action requests from unions in China, Korea, and Taiwan; translated them into English; and placed them on the electronic mail networks for international distribution. When necessary, the process was run in reverse.

International training and awareness programs in IT designed for the trade union movement need to be strengthened. The workers' education program of the International Labor Organization (ILO) recently established a number of international courses, as have several international trade unions for their own affiliates, not simply as technical training, but rather as a program of trade union education on the implications of telematics for the development of trade unionism itself.

Interunion discussion and agreement are needed concerning the potential for common online database provision and for drawing up detailed plans for information management. This would have to include a wide range of political discussions, but, in practical terms, could address two developments: the provision of electronic publishing facilities and the role of information brokers for the trade union movement.

Although the steps required to create online trade union databases are relatively very simple, there is a strong need for support and coordination, partly for technical reasons (e.g., ensuring that data are correctly formatted for the host computer). There is also a need for some form of overall editing and management, not to interfere with or amend the content, but to ensure that it is coherently organized within the database, which may contain information from dozens or hundreds of organizations in the labor movement.

Distinctions also have to be made between information placed online for widest possible public access and that limited to the broad trade union audience or to members or affiliates of specific organizations. Technically, this presents no problem, but it does require some level of collaboration and coordination to avoid having a morass of small individual databases, which can be bewildering to the individual user seeking specific information from a variety of sources.

A good information system also requires information brokers: individuals skilled at identifying and retrieving complex information from both labor movement databases and (more importantly, given the high costs involved) the hundreds of commercial, academic, and governmental databases. In effect, this is what research departments in the better resourced international and national unions already do, but as the number of unions online grows, the demand for these skills will accelerate.

Any strategy for the serious application of telematics to international trade unionism is going to take time, with considerable implications for organizational structure and political culture. Unions, despite popular mythology, are not monolithic hierarchies that can dictate change at will. They are complex and uniquely democratic institutions that do not adapt to new technologies quickly. Nevertheless, telematics technology does have inherently democratic features that, if harnessed, adapted, and extended by the international trade union movement, could revolutionize its ability to affect social change.

One of the most positive of recent developments has been the emergence of an education-based model of telematics to support international collective bargaining. Known as the International Study Circles, a program of transnational learning has been established by the International Federation of Workers' Education Associations (IFWEA)—working closely with a number of International Trade Secretariats.

The first full pilot ISC project was successfully completed in December 1997, studying the impact of transnational corporations in different workplaces and local communities across the world. Groups of workers and trade union representatives from Peru, Barbados, South Africa, Kenya, Estonia, Bulgaria, France, Belgium, Spain, Germany, Sweden, and the United Kingdom took part, working to a commonly agreed to set of materials, educational methodology, and so on. Over 100 representatives participated. A range of further ISC "courses" are planned, including "company-specific" courses.

The importance of this initiative is that it applies the technology to the best traditions of trade union education—participatory, democratic, student-centered, and so forth; that it takes an educational approach to the problems of information sharing and cooperation; and that it does not compromise the democratic autonomy of local/national trade union structures. Neither does it depend on individual internet access by each participant, but rather the technology is left outside the classroom and is used purely to assist the debates between the groups. (Details can be found at www.tsl.fi/isc.)

In the longer term, unhindered access to communications and information networks has to be regarded as a major prerequisite of international economic and social development, possessing a status similar to basic needs such as water and power. The labor movement, in partnership with other international social movements, has a potentially important role to play through inter-governmental agencies and international regulatory and industrial bodies (such as UNESCO and the International Telecommunications Union) to campaign for the rights of access to telecommunications networks, unimpeded and uncensored transborder communications, and electronic information media.

REFERENCES

Lane, T. (1987, February). Unions: Fit for active service? *Marxism Today*, pp. 20-21.

Levinson, C. (1972). *International trade unionism*. London: Allen & Unwin.

Marginson, P. (1992, December). European integration and transnational management-union relations in the enterprise. *British Journal of Industrial Relations*, pp. 529-545.

Mather, C., & Lowe, B. (1990). *Trade unions online: The international labour movement & computer communications*. Lancashire: Lancashire Polytechnic, Centre for Research on Employment and Work.

International Chemical and Energy Worker Federation (ICEF). (1990). *Report to the ICEF Executive Committee on ICEF research and the use of the research department as a source of information by affiliated trade unions*.

Towards effective responses to the challenge of internationalization. (1990, October). Paper presented to the Twenty-fifth Meeting of the ICFTU/ITS Working Party on Multinational Companies. Geneva, Switzerland.

About the Authors

EDITORS

Gerald Sussman is professor of political economy and communications in the School of Urban Studies and Planning and Department of Speech Communication, Portland State University. He has written widely on the political economy of Third World telecommunications and worked as a journalist, teacher, and researcher in Southeast Asia. His articles and chapters have appeared in popular and academic journals and in several books, including *The Political Economy of Information* (Ablex) and *Telecommunications Politics* (Lawrence Erlbaum). He edited (with John A. Lent) *Transnational Communications: Wiring the Third World* (Sage) and is the author of *Communication, Technology, and Politics in the Information Age* (Sage).

John A. Lent has taught mass communications since 1960. He has been a Fulbright scholar in the Philippines (1964-65), started and directed the first mass communications program in Malaysia (1972-74), taught at several universities in the United States, and has lectured at universities and conferences throughout the world. He has authored or edited 48 books and monographs and hundreds of articles on popular culture, Asian and Caribbean mass communications, cartooning, and critical studies in the mass media. His most recent books are *A Different Road Taken: Profiles in Critical Communication, Asian Popular Culture* and *Africa, Asia, Australia and Latin America: A Comprehensive International Bibliography.*

CONTRIBUTORS

James Cornford is a senior researcher at Newcastle University's Centre for Urban and Regional Development Studies (CURDS). His main interests concern the implications of technological and organizational change in the communications industries for the development of cities and regions. He is currently researching the emergence of the "information society" in European regions.

Vincent Mosco is professor of communication at Carleton University, Ottawa. His most recent book, *The Political Economy of Communication: Rethinking and Renewal* (Sage), funded under a five-year grant of the Canadian Social Sciences and Humanities Research Council, provides a political economy analysis of the philosophical, theoretical, and substantive areas of communications research, cultural studies and policy studies.

David F. Noble is a professor of social history of science and technology at York University. He is the author of *Forces of Production: A Social History of Industrial Automation* and *America By Design: Science, Technology, and the Rise of Corporate Capitalism.* His more recent books include *Progress Without People: The Politics of Technological Change, A World Without Women: The Christian Clerical Culture of Western Science,* and the forthcoming *The Religion of Technology.*

Manjunath Pendakur is professor of political economy of communications and chairs the Department of Radio-TV-Film at Northwestern University. His books include *Canadian Dreams and American Control: The Political Economy of the Canadian Film Industry* (Wayne State) and (co-edited) *Illuminating the Blindspots: Essays Honoring Dallas Smythe* (Ablex). He is currently writing a book on Indian popular cinema and television.

Charley Richardson is director of labor extension and heads the Technology and Work Program at the University of Massachusetts-Lowell, offering training, technical assistance, and strategic planning support to unions facing technological change, work reorganization, and new types of labor-management relations. He previously worked in shipyards as a shipfitter and as steward and safety observer for the IUMSWA Local 5 boilermakers union.

Kevin Robins is a professor of cultural geography in the Centre for Urban and Regional Development Studies, Newcastle University. His recent books include *Into the Image: Culture and Politics in the Field of Vision* (Routledge), and (with David Morley) *Spaces of Identity* (Routledge).

Sid Shniad is research director for the Telecommunications Workers Union in British Columbia and holds a Master's degree in labor economics. At TWU, he focuses on union strategies for long-distance competition, deregulation, and the information highway. Shniad is involved in deliberations on British Columbia's electronic highway accord.

Lenny Siegel is director of the Pacific Studies Center in Mountain View, CA, which monitors the U.S. military and high-technology industry, and is also director of Career/Pro at San Francisco State University. He edits the *Citizens' Report on the Military and the Environment* and, with John Markoff, wrote *The High Cost of High Tech* (Harper & Row).

Ewart Skinner is an assistant professor in the Department of Telecommunications at Bowling Green State University. A citizen of Trinidad and Tobago, Skinner has worked as a UNESCO project coordinator on the Caribbean mass media and as a media stringer in the Middle East. His research and publications focus on the area of information, communication, dependency, and culture in the Caribbean region.

David Spooner is a researcher, educator, and consultant in the United Kingdom. He has been the Principal Economic Development Officer for the Manchester City Council, an organizer of Manchester's Host and Electronic Village Hall projects, and London's Popular Telematics Project. He also has been a consultant to the British Labour Party and the United Nations International Labour Organization.

Lai Si Tsui-Auch completed her Ph.D. dissertation at Michigan State University on the regional division of production in the electronics industry between Hong Kong and Shenzhen Special Economic Zone. She is presently coordinating an international research group on organizational learning and conducting research on the downsizing of U.S., European, and Japanese firms in Europe at the Science Center of Social Research in Berlin.

Janet Wasko is a professor of communication at the University of Oregon. She is the author and editor of 10 books on the political economy of communications, including *Movies and Money* (Ablex), *Beyond the Silver Screen: Hollywood in the Information Age* (Polity), and *The Critical Communications Review* (3 volumes, edited with Vincent Mosco, Ablex).

Mark Wilson is associate professor at Michigan State University, James Madison College (chair, Political Economy) and the Institute for Public Policy and Social Research, where he directs the Electronic Space Project. His research is on the globalization of services and the spatial impacts of information technology at the urban and regional levels. He also directs the Nonprofit Michigan Project.

Author Index

Subject Index